Seismic Hazard and Building Vulnerability in Post-Soviet Central Asian Republics

NATO ASI Series

Advanced Science Institute Series

A Series presenting the results of activities sponsored by the NATO Science Committee, which aims at the dissemination of advanced scientific and technological knowledge, with a view to strengthening links between scientific communities.

The Series is published by an international board of publishers in conjunction with the NATO Scientific Affairs Division

A Life Sciences **B Physics**	Plenum Publishing Corporation London and New York
C Mathematical and Physical Sciences **D Behavioural and Social Sciences** **E Applied Sciences**	Kluwer Academic Publishers Dordrecht, Boston and London
F Computer and Systems Sciences **G Ecological Sciences** **H Cell Biology** **I Global Environment Change**	Springer-Verlag Berlin, Heidelberg, New York, London, Paris and Tokyo

PARTNERSHIP SUB-SERIES

1. Disarmament Technologies	Kluwer Academic Publishers
2. Environment	Springer-Verlag / Kluwer Academic Publishers
3. High Technology	Kluwer Academic Publishers
4. Science and Technology Policy	Kluwer Academic Publishers
5. Computer Networking	Kluwer Academic Publishers

The Partnership Sub-Series incorporates activities undertaken in collaboration with NATO's Cooperation Partners, the countries of the CIS and Central and Eastern Europe, in Priority Areas of concern to those countries.

NATO-PCO-DATA BASE

The electronic index to the NATO ASI Series provides full bibliographical references (with keywords and/or abstracts) to about 50,000 contributions from international scientists published in all sections of the NATO ASI Series. Access to the NATO-PCO-DATA BASE is possible via a CD-ROM "NATO Science and Technology Disk" with user-friendly retrieval software in English, French, and German (©WTV GmbH and DATAWARE Technologies, Inc. 1989). The CD-ROM contains the AGARD Aerospace Database.

The CD-ROM can be ordered through any member of the Board of Publishers or through NATO-PCO, Overijse, Belgium.

2. Environment – Vol. 52

Seismic Hazard and Building Vulnerability in Post-Soviet Central Asian Republics

edited by

Stephanie A. King
Stanford University,
Stanford, California, U.S.A.

Vitaly I. Khalturin
United Institute of the Physics of the Earth,
Moscow, Russia

and

Brian E. Tucker
GeoHazards International,
Palo Alto, California, U.S.A.

Springer-Science+Business Media, B.V.

Proceedings of the NATO Advanced Research Workshop on
Earthquake Risk Management Strategies for Post-Soviet Central Asian Republics:
Avoiding Repetition of 1988 A Sakhalin Disasters
Almaty, Kazakhstan
22–25 October 1996

A C.I.P. Catalogue record for this book is available from the Library of Congress.

ISBN 978-90-481-5176-9 ISBN 978-94-017-2971-0 (eBook)
DOI 10.1007/978-94-017-2971-0

Printed on acid-free paper

TABLE OF CONTENTS

PREFACE

The dissolution of the Soviet Union in 1991 triggered a renaissance bordering on chaos in the five Central Asian republics of Kazakhstan, Kyrgyzstan, Tajikistan, Turkmenistan, and Uzbekistan. After 70 years of technical, political, and economic dependence on Moscow, these republics were, instantaneously and (to a certain extent) against their wishes, on their own. Each had to form a new system of government. None could guarantee its territorial security, even though one, Kazakhstan, found itself the world's fourth-largest nuclear power. A bloody civil war broke out in Tajikistan that threatened to spread to its neighbors. Each republic established its own official state language, replacing Russian. The region's ruble-based economy was abandoned in favor of five new and noninterchangeable currencies. Inflation soared. Millions of people of non-Central Asian heritage emigrated. Religion burst onto the scene: In 1989, each capital city averaged 10 mosques; two years later, this number had grown to several hundred. In place of security and stability came vulnerability and volatility, and most importantly, opportunity.

Given the enormity of the challenges they faced, these republics have made and continue to make remarkable progress. They have held elections and developed foreign policies. The civil war in Tajikistan has not spread and shows signs of ending. Inflation has abated. Investors have been attracted to the region's natural resources, which include some of the largest deposits of minerals, oil, and natural gas in the world. Billions of dollars in foreign currency are being spent on oil and gas exploration, automobile factories, telecommunication networks, international airports, and hotels. Pipelines are planned to stretch to the China Sea, the Indian Ocean, and the Mediterranean. A trans-Asian railroad and highway are under construction, and will connect the republics to each other and their immediate neighbors. These enterprises are forging new commercial and cultural links between Central Asia and the rest of the world, accelerating the region's political, social, and economic development. If Central Asia can survive these transitional years, its future is bright indeed.

One of the threats to Central Asia's future development is the region's large and growing urban earthquake risk.

Central Asia's earthquake activity has long been recognized as one of the highest in the world, but the extreme vulnerability of its Soviet-era residential buildings was realized only after two recent earthquakes outside the region. In 1988, an earthquake in Armenia caused the collapse of more than 95% of one type of residential building and 75% of another type in the city of Leninakan; other types of buildings in that city remained standing but were damaged. In 1995, another earthquake near Sakhalin, an island in the northwest Pacific Ocean, caused all of yet another type of residential building to collapse in the city of Neftegorsk; again, other building types survived. These experiences in Armenia and Sakhalin suggest that the thousands of residential buildings with similar design and construction found throughout Central Asia are highly vulnerable to earthquakes.

Just as Central Asia's large urban earthquake risk was being recognized, the ability to manage it was drastically decreasing. Since the Soviet Union's disintegration, responsibility for earthquake preparedness and response has been turned over to local officials, who are often inexperienced and usually more than occupied with present day emergencies. None of the five republics has a standing army capable of managing the consequences of a natural catastrophe. Among the millions of people who recently emigrated were about half of Central Asia's most experienced civil engineers and earth scientists. Those who remain are isolated from their colleagues in other republics and have difficulty attracting students to their professions. Funding for research and development has virtually ceased. For all of these reasons, it is understandable that the lessons of Armenia and Sakhalin have gone unheeded. But continuing to ignore them is unacceptable for both Central Asians, who live there, and the world community, which is poised to pour additional investments into the region.

Recognizing the urgency of addressing Central Asia's urban earthquake risk, GeoHazards International organized a NATO Advanced Research Workshop to assess the vulnerability of the region's Soviet-era residential buildings and develop a strategy for reducing it. The government of Kazakhstan agreed to act as host.

Support for organizing this workshop came from a wide variety of organizations. The initial seed funding came from NATO's Scientific and Environmental Affairs Division. Additional, essential financial support came from (listed in alphabetical order): the Foreign Office of the Federal Republic of Germany, GeoHazards International, the Kazakh State Committee for Emergencies, the United Nations University and the US Geological Survey. Other important support was provided by the Applied Technology Council (USA); the Cecil and Ida Green Foundation (USA); the German Association of Earthquake Engineering and Structural Dynamics; the International Association of Earthquake Engineering's World Seismic Safety Initiative; the International Association of Seismology and Physics of the Earth's Interior; the IRIS Consortium (USA); the Joint Seismic Program of Lamont-Doherty Geological Observatory of Columbia University (USA); the Kazakh Research and Experimental Design Institute on Earthquake Engineering and Architecture; OYO Corporation (Japan); the United Nations Educational, Scientific, and Cultural Organization (UNESCO); and the US National Center for Earthquake Engineering. The editors wish to express their gratitude to all these organizations, whose contributions made the workshop a success.

The resulting workshop was held in Almaty, Kazakhstan, from October 22-25, 1996, and involved more than 50 experts from the fields of seismology, earthquake-resistant design, and emergency response from across Central Asia and around the world.

This volume contains papers that were prepared for presentation and discussion at the Almaty workshop. Following the Executive Summary, which summarizes the outcome of the workshop, the next two papers provide an overview of the seismic hazard and building vulnerability, respectively, in the Central Asian republics. The next five papers are reports on seismic hazard and building vulnerability in each of the five Central Asian republics prepared by the workshop participants from each republic prior to the meeting in Almaty. These papers are based on responses to a series of questions pertaining to seismic hazard and building vulnerability that were formulated by the conference organizers. The questions, included in this volume in the Appendix, were designed to help the experts in each republic prepare comparable reports that were made available at the time of the Almaty workshop.

The next three papers describe observations and analysis of building damage in the 1988 Spitak, Armenia earthquake, the 1994 Kuril Islands earthquake, and the 1995 Sakhalin earthquake. Many of the buildings destroyed in these earthquakes are of similar design and construction to buildings located in the Central Asian republics. The final paper is a study of the seismic resistance of mass-constructed Soviet-era buildings that are located throughout Central Asia, using the city of Almaty as an example.

The editors wish to express, on behalf of all the participants of the Almaty workshop, their deep appreciation to several individuals whose personal efforts made this workshop and, therefore, this book possible. The Honorable Nikolay Makievsky, Deputy Prime Minister of Kazakhstan, provided the local overall support and hospitality that allowed the workshop to take place. The keynote speeches by him and by the Honorable Elizabeth Jones, Ambassador of the US to Kazakhstan, and by the Honorable Henning von Wistinghausen, Ambassador of the Federal Republic of Germany, underlined the need for the workshop and motivated the participants in their work. Academician Toeleby Zhunusov, Director of the Kazakh Research and Experimental Design Institute of Aseismic Engineering and Architecture, made available the resources of his institute. This workshop was triggered by a paper of William Leith, in June 1995, in which he pointed out that the consequences of the Sakhalin Earthquake should renew concerns about seismic safety in the former Soviet Union; he provided encouragement and resources throughout the organization of the workshop. Günter Klein and Christopher Rojahn provided almost daily support and advice while the workshop was organized and conducted. The concern for Central Asia and the technical expertise of all the workshop participants – largely unnamed in this book – shaped in very real ways the eleven papers presented here; many of these participants carefully prepared for the workshop and traveled long distances to attend. Dr. Luis Veiga Da Cunha and Alison Trapp of the Scientific and Environmental Affairs Division of NATO patiently guided us throughout the entire process of applying for support up to and including conducting the workshop itself. Wil Bruins and Annelies Kersbergen of Kluwer Academic Publishers assisted us in the publication of the manuscript. Finally, the person most responsible for the multitude of logistical arrangements of the workshop and without whose help the workshop would not have been a success, is Cheryl Eichorn, of the U.S. Geological Survey.

In closing we would like to urge readers to consider how best to help the peoples of Central Asia. As mentioned already, Central Asia has experienced for centuries the severest social, political, religious, and economic changes. These changes continue to this day. While the opportunity now exists for social stability, political freedom and economic development, these have not yet been completely achieved. At such a time, it may seem ill-considered to draw attention to yet another problem – urban earthquake risk, especially one that will occur at some unknown time in the future, with some unknown consequences. Why not let the Central Asians alone to deal with today's challenges?

For us, the question is not "either – or". The answer is that the Central Asians and international developers should face today's challenges with the *inevitable* large, future earthquakes in mind. When investing in infrastructure, developers should insist on employing seismically-resistant design and construction methods. When devising legal and political reforms, public officials should consider the need to create, maintain and enforce modern building codes. When expanding the freedom of the press and other media, leaders should be aware of the need to inform honestly the public of the risk involved in living and working in the many seismically-vulnerable structures built during the Soviet era. Failing to take into account Central Asia's earthquake risk puts all the current and future development and social progress in jeopardy. We hope that this book contributes in a small way to the rapid development of Central Asia and to the safety of its people.

S. KING
Stanford University
Stanford, CA, USA

V. KHALTURIN
United Institute of the
Physics of the Earth
Moscow, Russia

B. TUCKER
GeoHazards International
Palo Alto, CA, USA

EXECUTIVE SUMMARY AND WORKSHOP SUMMARY

1. Central Asia's Earthquake Hazard

The earthquake hazard, in terms of the maximum ground shaking expected at a given location over a specified period of time, of the most populated portion of Central Asia is approximately equal to that of California. More than 50 million people live in the Central Asian republics (see Table 1 for a summary of the region's demographics). Central Asia's earthquakes and two-thirds of its population are concentrated in the region's southern quarter, which has about twice California's area and about twice its number of annual earthquakes.

TABLE 1. Central Asia demographics

Republics			Capital Cities	
Name	Population (millions)	Area (x1000 sq. km.)	Name	Population (millions)
Kazakhstan	17.0	2,720	Almaty	1.5
Kyrgyzstan	4.4	200	Bishkek	0.8
Tajikistan	5.8	140	Dushanbe	1.1
Turkmenistan	4.5	490	Ashgabad	0.5
Uzbekistan	22.7	450	Tashkent	2.2
All republics	54.4	4,000	All capitals	6.1
Contiguous United States	260.0	7,884		

Earthquake hazard is often expressed in terms of seismic intensity, which is a qualitative description of the consequences of earthquake shaking on people and structures. In the former Soviet Union, seismic intensity is measured on a 12-step scale, called the Medvedev-Sponheuer-Kárník (MSK) scale. This scale, a portion of which is shown in Table 2, is similar to the Modified Mercalli Intensity (MMI) scale used in the United States and Europe.

TABLE 2. Partial definition of MSK intensity scale

MSK Intensity	Consequences	
	People	Buildings
VII	Frightened	Poor-quality structures considerably damaged; ordinary structures slightly damaged
VIII	General fright, some panic, difficulty standing	Poor-quality structures collapsed; ordinary structures considerably damaged; and well-built structures slightly damaged
IX	General panic	Many ordinary structures destroyed; well-built structures heavily damaged
X	Thrown to ground, strong disorientation	Most buildings destroyed, including some well-built structures

Maps of seismic hazard in Central Asia have been derived primarily from descriptions of the consequences of past earthquakes. These records show that, over the last century alone, all of the region's capitals were heavily damaged by earthquakes and some were totally destroyed; for example, Ashgabad in 1948, and Almaty in 1887 and again in 1911.

The simplified version of the official seismic hazard map for the former Soviet Union that is shown in Figure 1 indicates that all of the Central Asian capitals, with the exception of Tashkent, can expect an MSK IX level of shaking. Tashkent can expect MSK VIII. The period of time over which this level of shaking is expected varies from location to location as described in the paper summarizing the seismic hazard in Central Asia.

There are two reasons to believe that this official map significantly underestimates the region's hazard. First, it does not take into account the amplification effect of the soft-soil conditions common in large areas of the capitals, which is important because soft soils can produce intensities one or more MSK units greater than on nearby stiff soils. Second, as shown in Table 3, almost all of the recent destructive earthquakes in the former Soviet Union have been significantly larger than would be expected from examining the map (even allowing for soft-soil conditions). This underestimation of seismic hazard is partially (but only partially) responsible for the widespread collapse of buildings in Armenia and Sakhalin, because those structures were designed to withstand smaller ground motions than actually occurred. This map is currently being revised in Moscow.

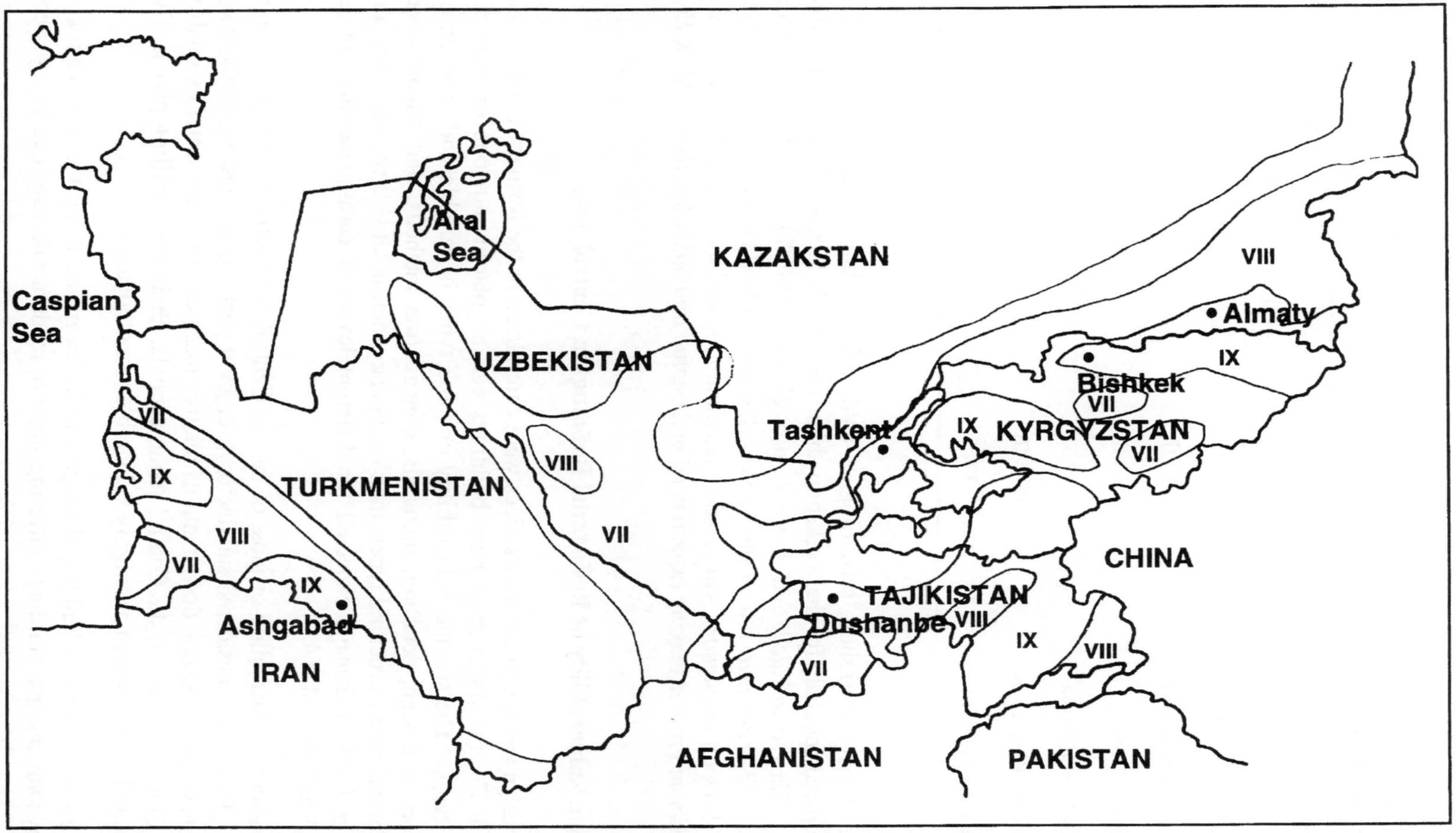

Figure 1. Simplified version of the 1978 official Soviet seismic hazard map showing maximum expected seismic intensity (MSK) in Central Asia.

TABLE 3. Comparison of maximum intensity expected from the 1978 Soviet Seismic Hazard Map and maximum intensity observed for destructive earthquakes since 1988

Earthquake	Maximum Intensity Expected (MSK)	Maximum Intensity Observed (MSK)
1988 Leninakan, Armenia	VIII	IX
1990 Zaysan, Kazakhstan	VI-VII	VIII
1991 Racha, Georgia	VII-VIII	IX
1991 East Siberia, Russia	V	VIII-IX
1992 Soosamir, Kyrgyzstan	VII-VIII	IX
1995 Neftegorsk, Russia	VII	IX

At the Almaty workshop, seismologists analyzed the above information and the reports prepared specially for the workshop about the seismic hazard of each of the five republics. *They concluded that there is a high (about 40%) probability that an earthquake will occur near one of the Central Asian republics' capitals within the next 20 years. Such an earthquake will produce maximum ground shaking in that city equal to the maximum ground shaking experienced in Armenia and Sakhalin, that is, MSK IX.*

2. Seismic Vulnerability of Residential Buildings in Central Asia

Because design and construction practices were centralized in the former Soviet Union, 80% of all Central Asian residential buildings can be placed into one of only six structural types. The seismic vulnerability of these types is variable and depends on such factors as design, detailing, materials, construction methods, and maintenance. The six Central Asian structural types, their occupancy total in all five capital cities, and the average level of damage expected for different levels of earthquake shaking are described briefly in Table 4.

The seismic vulnerability of most of the six Central Asian structural types is high. Only one (Type 6) is considered satisfactory; its good performance during earthquakes is due to its seismic-resistant design and its relative insensitivity to construction quality. One-half of the residents of the Central Asian capitals, about three million people, live in buildings (Types 1-5) that are highly vulnerable to earthquakes.

The economic cost of building damage can be estimated using the information in Table 4 by knowing that buildings suffering slight or moderate damage can be repaired,

TABLE 4. Central Asian structural types, their occupancy total in all five capital cities, and the expected damage levels

Structural Type	Occupancy		Damage Level		
	Thousands	% Urban Population	MSK VII	MSK VIII	MSK IX
1. Non-engineered structures, including small adobe and unreinforced masonry buildings	1,200	20%	Heavy damage	Partial to total collapse	Total collapse
2. Brick, bearing-wall systems with wooden floors, one to two stories, pre-1955			Moderate to heavy damage	Partial collapse	Total collapse
3. Brick, bearing-wall systems with precast reinforced concrete floors, three to five stories, pre-1957	1,400	23%	Slight to moderate damage	Heavy damage to partial collapse	Partial collapse
4. Brick, bearing-wall systems with precast reinforced concrete floors, some seismic detailing, post-1957			No damage to slight damage	Moderate to heavy damage	Heavy damage to partial collapse
5. Precast reinforced concrete frames with welded joints and brick infill walls, four to nine stories	400	7%	Slight damage	Moderate to heavy damage	Heavy damage to partial collapse
6. Precast reinforced concrete large-panel systems with dry or wet joints	1,800	30%	No damage to slight damage	Slight to moderate damage	Moderate damage
Other	1,300	20%	---	---	---
Total	6,100	100%			

buildings suffering heavy damage might be repaired, and buildings that partially or completely collapse cannot be repaired.

Building damage also has a human cost. Based on worldwide experience, it is estimated that the fatality rate in urban centers of developing countries will be 0.5% for MSK VIII and 5% to 7% for MSK IX. Similarly, it is estimated that the rate of serious injuries (i.e., those requiring hospitalization) will be 2% for MSK VIII and 20% for MSK IX. The expected number of deaths and injuries the Central Asian capitals can be estimated assuming MSK IX intensity in Almaty, Ashgabad, Bishkek, and Dushanbe; and assuming MSK VIII intensity in the 60% of Tashkent's area that has stiff soil conditions and MSK IX in the 40% with soft-soil conditions (see Table 5).

TABLE 5. Estimated deaths and injuries in Central Asian capitals

City, Republic	Serious Injuries (Thousands)	Deaths (Thousands)
Almaty, Kazakhstan	300	75
Ashgabad, Turkmenistan	100	25
Bishkek, Kyrgyzstan	160	40
Dushanbe, Tajikistan	220	55
Tashkent, Uzbekistan	180	45

At the Almaty workshop, structural engineers analyzed this information and the reports specially prepared for the workshop about each of the five capitals' building stock. *They concluded that it should be expected that an MSK IX level of ground shaking in a Central Asian capital will cause tens of thousands of fatalities, and at least a hundred thousand serious injuries. As many as half of the city's residential buildings will collapse or be damaged beyond repair.*

3. A Call to Action

Central Asia's urban earthquake risk is unusually easy to evaluate. Its buildings vary little in design and method of construction because the vast majority of them were built over a short period, when design and construction were controlled by one central authority. Further, how some of these building types perform in earthquakes has been tested and found to be poor, first in Armenia and again in Sakhalin.

Consequently, the earthquake specialists who gathered from across Central Asia and around the world at the Almaty workshop could agree that there is a high probability that, during the next several decades, a large earthquake near one of the

Central Asian capitals will cause human and economic loss even greater than that already experienced in Armenia and Sakhalin … unless corrective action is taken soon.

The workshop participants concluded that, in order to confront this crisis, projects must immediately be initiated that allow for Central Asia's current social, political, and economic conditions, and address the following five broad needs:

1. *Inform the people most at risk.* Responsible officials in each republic must first notify the occupants of Soviet-era residential buildings of the high vulnerability of some of these buildings, and next undertake a detailed inventory and ranking of vulnerable buildings in their respective capitals. It is a basic human right to know if one is exposing oneself and one's family to great risk. Informing those who are at great risk would be not only a responsible but also an effective first step, because projects to improve seismic safety in Central Asia are possible today only with the strong support of the public.

2. *Rehabilitate existing buildings.* A seismic rehabilitation program should be launched in the capital of each republic to upgrade all highly vulnerable multifamily residential structures. The uniformity of Soviet-era construction makes rehabilitation uniquely practical. While this program is being planned, a demonstration and training project might be conducted on, for example, a foreign embassy or foreign office building, for which the necessary funding could quickly be made available.

3. *Regulate new construction.* New seismic design codes should be written taking into account currently available material and construction methods. Designs that minimize sensitivity to construction quality, such as that of structural Type 6, are desirable. Liability for illegal construction must be established. Sharing the experience of other nations in drafting, enforcing, and updating seismic safety laws would be fruitful. New construction must be continuously inspected by trained and independent public officials, who can be held accountable. Lethal construction must cease.

4. *Unite and support local experts.* Central Asia's too few, underfunded, and isolated earthquake engineers and seismologists must reestablish contact with each other and create new links with international colleagues, including recent émigrés. Exchange of information will help the republics to train new professionals, establish laws and standards, and advocate earthquake safety. Collaboration should be increased with Internet connections, attendance at international conferences, subscriptions to foreign professional journals, and cooperative research projects.

5. *Continue and extend risk assessment.* Estimates of earthquake risk based on seismic intensity records are not adequate to design public policy. A network of strong-motion accelerometers across each capital city and in standard buildings would determine local ground response and building performance. Maps of soil conditions would also be useful. Finally, while the Almaty workshop focused on residential buildings, it also revealed that the earthquake resistance of other

structures is highly suspect. Consequently, an assessment of the seismic vulnerability of critical structures such as schools, hospitals, government buildings, and lifelines should immediately be undertaken.

The participants of the Almaty workshop assessed the earthquake risk of Central Asia's Soviet-era residential buildings and recommended means to manage it. Now is the time for others to act who understand the risk faced by their families and their communities. Only a group of concerned, determined Central Asian citizens – from the very highest government officials to civil servants, parents, and teachers – can take the actions required to avert tragedy.

SEISMIC HAZARD OF THE CENTRAL ASIA REGION

A. NURMAGAMBETOV
UIPE, Complex Seismological Expedition, Kazakhstan
N. MIKHAILOVA
KazNIISSA, Kazakhstan
W. IWAN
California Institute of Technology, USA

1. Introduction

The territory under review is the north part of Central Asia including five republics of the former Soviet Union – Kazakhstan, Uzbekistan, Kyrgyzstan, Tajikistan, and Turkmenistan. It is a very complicated region in its geological-tectonic aspect, and it is at present one of the most highly seismic geostructural areas in the world. Much research work in various scientific fields has been directed towards studying the nature of earthquakes, the assessment of seismic hazard, and the development of methodologies for forecasting large earthquakes. A large amount of material has been collected on the different aspects of geology, tectonics, and seismic activity of the region, which shows the high level of seismic hazard in most parts of the republics, including the capital cities.

This chapter gives the characteristics of the seismic and seismic-tectonic conditions of the region, including the general approach and results of seismic zoning and assessment of seismic hazard in the capital cities. This is intended as a review to attract attention to the problems of seismic risk mitigation and the reduction of possible future earthquake damage in Central Asia.

The material in this chapter is based on reports made by the participants at the International NATO Advanced Research Workshop on "Strategies for Seismic Risk Reduction in Urban Territories of Central Asia" held in October of 1996 [1-5]. These reports are included in this book.

2. Seismic Intensity in the Former Soviet Union

Seismic hazard may be defined as the probability that an earthquake of a specified size will occur in a given region within a specified interval of time. Implicit in this definition is the specification of the size of the earthquake. The size of an earthquake

S.A. King et al. (eds.),
Seismic Hazard and Building Vulnerability in Post-Soviet Central Asian Republics, 1–43.

may be specified in many ways, ranging from estimated energy release, measures of ground shaking, or statements about the consequences of the earthquake. Historically, much effort has been directed toward defining measures of the size of an earthquake that are related to its effects on man-made structures.

Attempts to quantify the size of an earthquake based on its consequences date back to the 17th century. In this approach, the size of an earthquake is commonly classified according to a scale of intensity. One of the earliest intensity scales was that published by Mercalli in 1897. In 1917, this scale was modified by Cancani and Sieberg, and became known as the MCS scale. This scale is still in use in some European countries. In 1931, Wood and Neumann suggested further revisions of the Mercalli scale, and the resulting twelve step intensity scale became known as the Modified Mercalli Intensity (MMI) scale. This scale is still used in the United States.

In 1952, the Soviet Academy of Sciences proposed a twelve step scale to describe the consequences of earthquakes in the Soviet Union. This scale was quite similar to the MCS scale. It was used until 1964 when it was refined by Medvedev, Sponheuer, and Kárník. The resulting intensity scale is referred to as the MSK-64 scale. It is the scale currently employed in the former Soviet Union. The MSK scale is very similar to the MMI scale. According to the official text of the MSK-64 scale [29], some of the consequences of earthquakes of intensities VI-X are as indicated in Table 1.

TABLE 1. Consequences of earthquakes of intensities VI through X, from official text of MSK-64 scale [29]

MSK Intensity	Title and Description of Damage
VI	*Fright.* Slight damage to many adobe buildings and buildings made of broken stones; slight damage to individual buildings made of large blocks and panels, and frame structures
VII	*Damage to buildings.* Moderate damage to many buildings made of large blocks and panels, and frame structures; slight damage to many reinforced concrete frame buildings
VIII	*Heavy damage to buildings.* Heavy damage and occasional destruction of buildings made of large blocks and panels; moderate damage to many reinforced concrete frame buildings
IX	*Partial destruction of buildings.* Partial destruction and occasional collapse of buildings made of large blocks and panels; partial destruction of reinforced concrete frame buildings
X	*Total destruction of buildings.* Collapse of many buildings made of large blocks and panels, and frame structures; partial destruction and occasional collapse of reinforced concrete buildings

3. Characteristics of Seismic-tectonic Conditions in Central Asia

The territory of Central Asia consists of high mountains and significantly fragmented geologic structures [6]. Currently, the geology consists of: Turan segment of young (epipaleozoic) platform, alpine mountain-folded structures of Kopetdag and Pamirs, and platform orogenic areas of Tien-Shan and Djungaria (see Figure 1). Each of these regions is a fragment of even larger zones of tectonically similar structures covering the majority of the European-Asian continent [7].

There are a number of geologic hypotheses that attempt to explain the specific structure and endogene activity of this territory. The most widely accepted hypothesis if that of the Indostan plate producing pressure on the European-Asian continent [8]. According to this theory, the Indostan plate not only produces pressure on the European-Asian continent, it also moves under the continent, which explains the high concentration of mountain ranges in Central Asia.

The earth crust of this region is broken up by a system of abyssal faults, which are natural boundaries of the largest geologic blocks [9-11,26] (see Figure 2). The faults are characterized by a wide dispersion of directions – from sublatitude to submeridianal (north-west and north-east are the prevailing directions). Almost all zones of abyssal breaches are of Paleozoic age. The faults differ considerably in their level of seismicity. In addition, seismic activity has changed considerably over time, which can be seen from the results of the latest paleoseismic dislocations research and observations of large and small earthquake movements.

The seismic history of the territory is rich. Since ancient times, various sources have revealed numerous catastrophic earthquakes [12]. During the past approximately 100 years, four earthquakes of magnitude greater than 8 have occurred: Krasnovodsk earthquake of 1895, Kashgar earthquake of 1902, Chilik earthquake of 1889, and Kemin earthquake of 1911. Figure 3 shows a map of the epicenters of large crustal earthquakes (magnitude greater than 6.0) for the past 150 years. The map helps to point out seismic generating structures that may not be evident from geology alone.

In general, the location of large earthquakes is dependent on the geometry of long existing faults. When comparing the seismic intensity of the earthquakes with the tectonic situation, it is possible to single out zones of high seismicity: Pamirs-Alai, Gissar-Karakul, East-Fergana, Chatkal, North Tien-Shan and also Pamirs-Hindukusk zones of deep-focus epicenters. A number of less considerable zones can also be singled out.

Numerous residual deformations in the region can be revealed that show high seismic activity in the distant past, as researchers do not consider the deformations to be connected with any earthquakes in the recent past [9]. Figure 4 shows the epicenters of assumed earthquakes based on paleoseismic geologic data. The distinctive configurations have zones of dislocation coinciding with the Talass-Fergana, Chilik-Kemin, and Darvaz-Karakul fault zones.

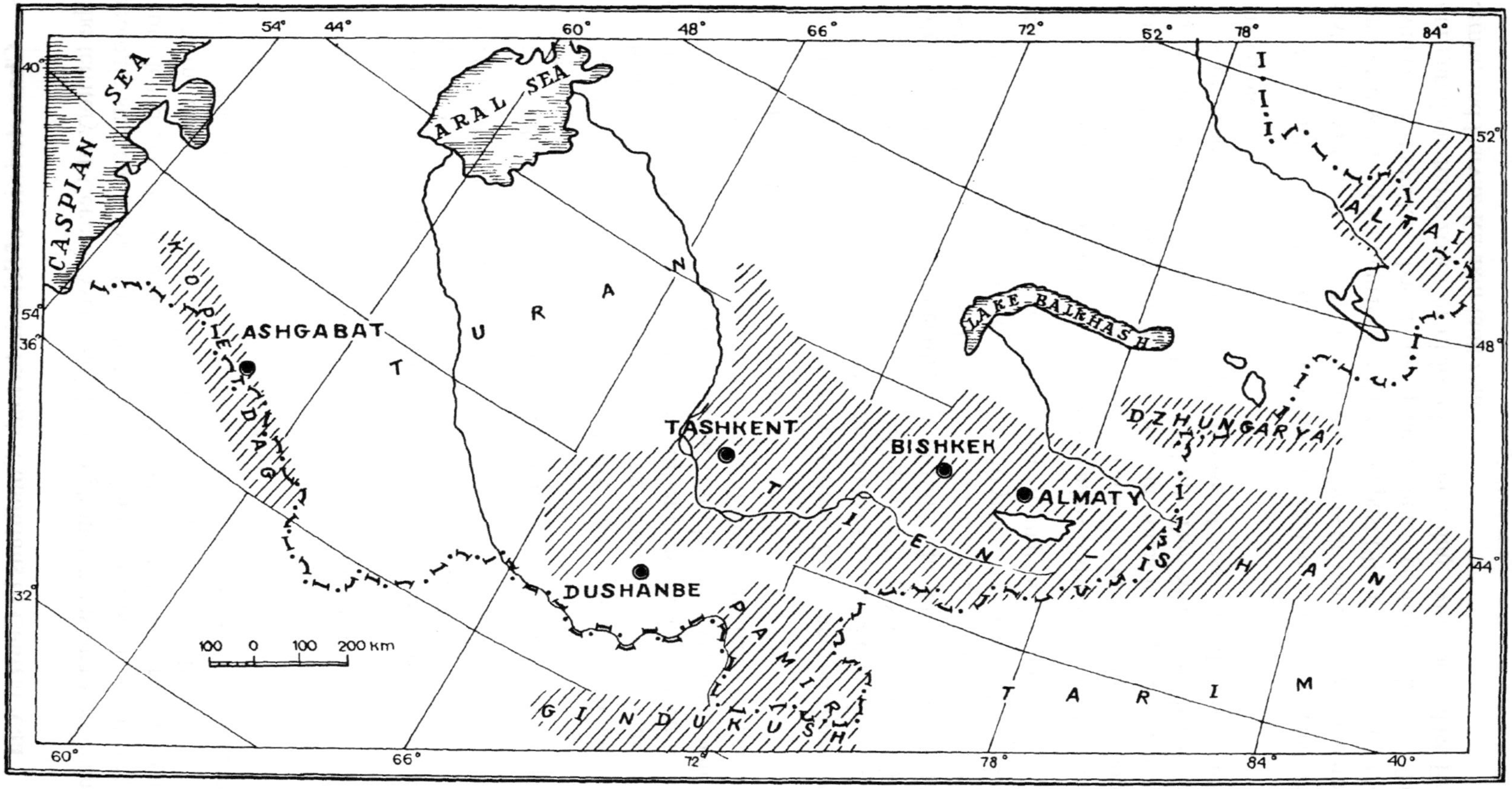

Figure 1. Map of the Central Asian region.

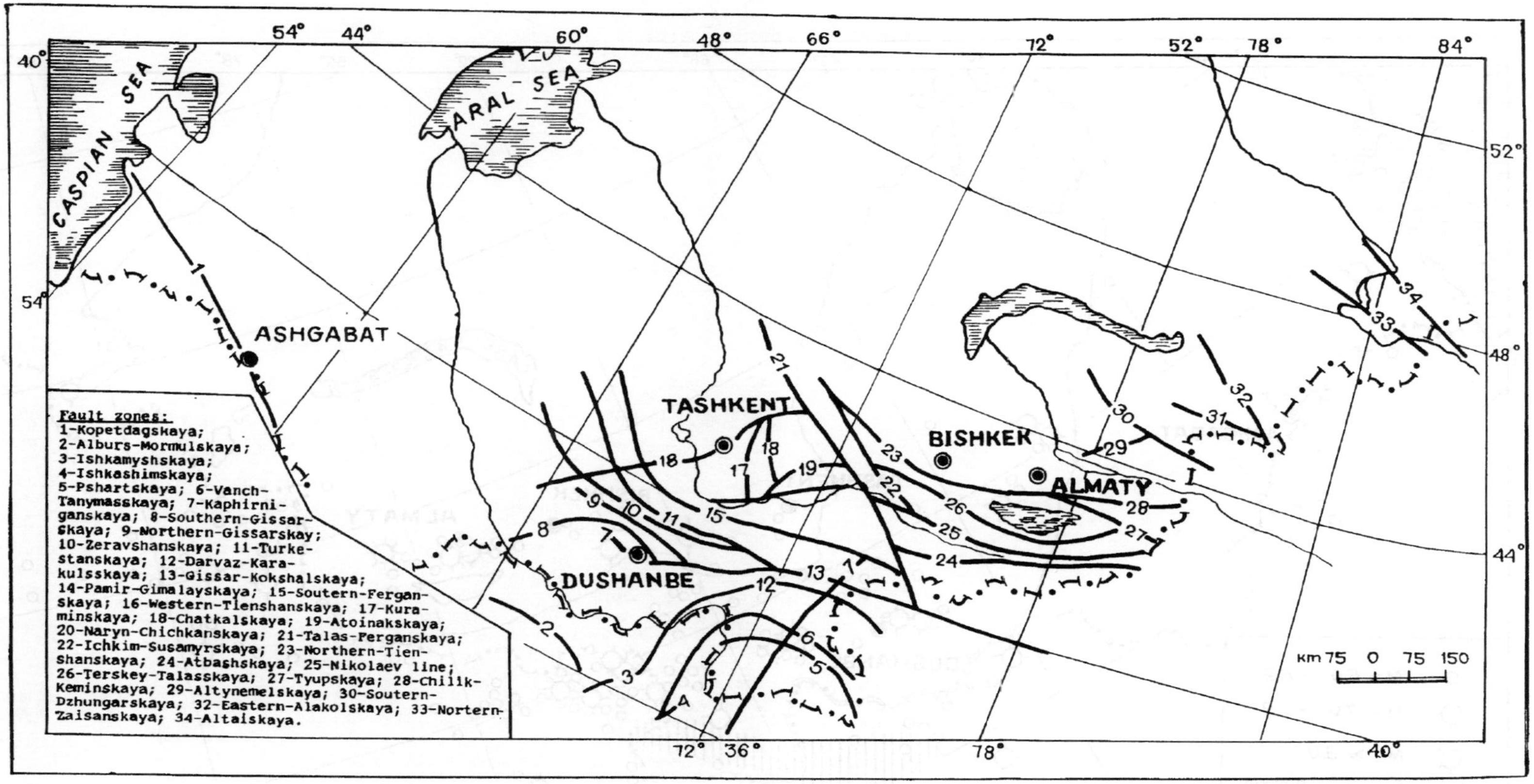

Figure 2. Zones of significant deep faults [9-11, 26]

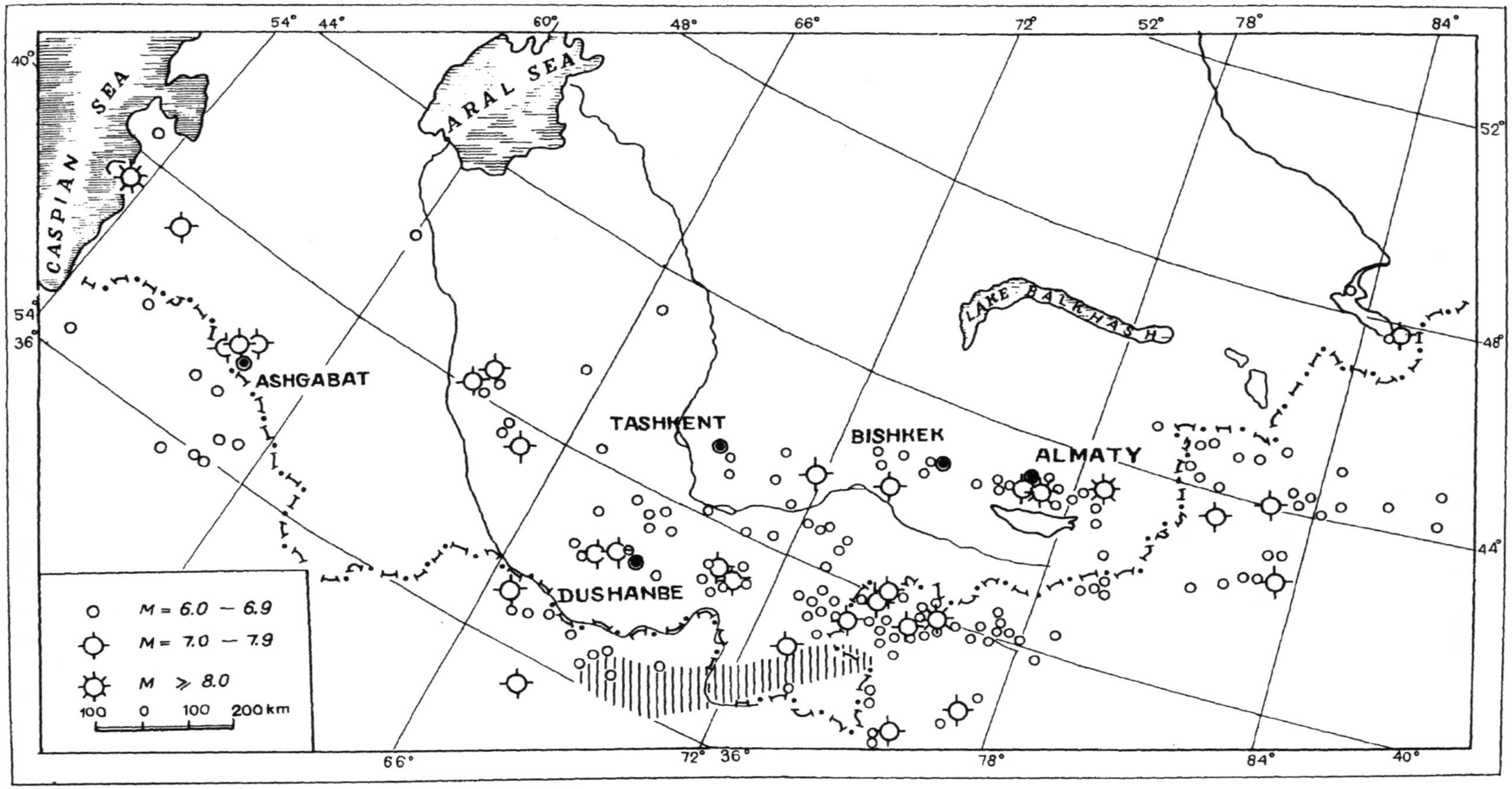

Figure 3. Map of epicenters of strong shallow earthquakes. Area of epicenters of deep Pamir and Hindukusk earthquakes is shaded.

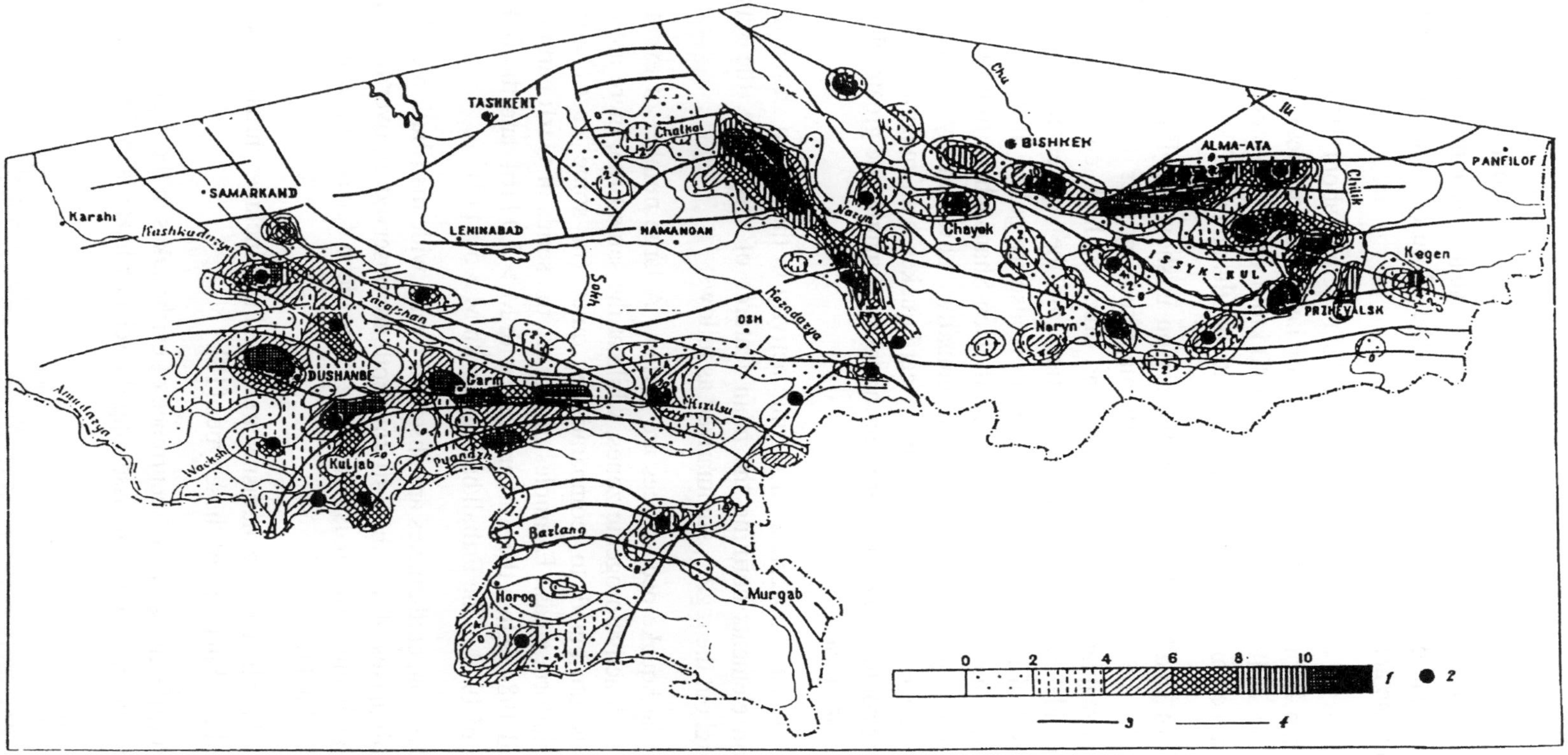

Figure 4. Epicenters of earthquakes according to paleoseismic geological data [9]. 1 = scale of deformation density; 2 = earthquake epicenters or M ≥ 6.5 obtained from paleoseismic data; 3 = fault zones; and 4 = regional faults with an anti-Tien-Shan strike.

The current seismic activity in the region is observed by more than 130 seismic recording stations, the first of which was established at the start of the 20th century. Each Central Asian republic has its own network of stations. Until 1990, the number of stations was increasing steadily, however, in the past few years, the number of stations has decreased. Financial difficulties have lead to the closure of several stations, especially those in remote areas.

The annual catalogs of earthquakes in Central Asia include about 3000 events of magnitude greater than 2.5 (excluding aftershocks of large earthquakes). About one half of these events are shallow (in the crust, up to 40 km deep). The others are under the crust, primarily in the Pamirs-Hindukush zone with depths up to 270 km.

The distribution of small earthquakes in the region is shown in Figure 5, which includes all events in the past 30 years with magnitude greater than 3.0. This distribution is very irregular. There are lines and areas of high concentrations of epicenters that coincide with the areas identified by the epicenters of strong earthquakes. The level of moderate seismic activity is also irregular. It is highest for the Pamirs-Hindukush zone and the ranges of south Tien-Shan, where more than 100 earthquakes of magnitude greater than 3.0 occur annually. Low activity is typical for the north Tien-Shan ranges where the number of earthquakes with magnitude greater than 3.0 is on the order of ten; however, large earthquakes of magnitude greater than 8.0 have occurred here in the past.

Analysis of the available materials on seismic activity in the region shows that in spite of the common elements defining the main features of seismicity, the seismic processes vary throughout the region, summarized as follows:

1. Sources of earthquake occurrences are varied. A large number of events are connected with active orogenic zones, while at the same time, there are rather large events attributed to comparatively quiet (in some cases assumed aseismic) platform areas. Examples of this are the series of Gazli earthquakes of 1976 and 1984 within the Turkan plate with $M > 7.0$ [16], and the Bakanas earthquake of 1979 in the Pribalkhask gap with $M > 5.8$ [17].

2. Within the region, earthquakes are observed over a wide distribution of depths. Crustal earthquakes of less than 40 km depth are observed all over the region, while in the Pamirs-Hindukush zone there are deep focus events located at depths up to 300 km.

3. The types of earthquake mechanisms are varied and include: thrust, strike-slip, normal, and combinations of these types [18].

Such significant differences in the occurrence of earthquakes in the region show the difficulty in defining the seismicity and assessing the seismic hazards in the republics of Central Asia.

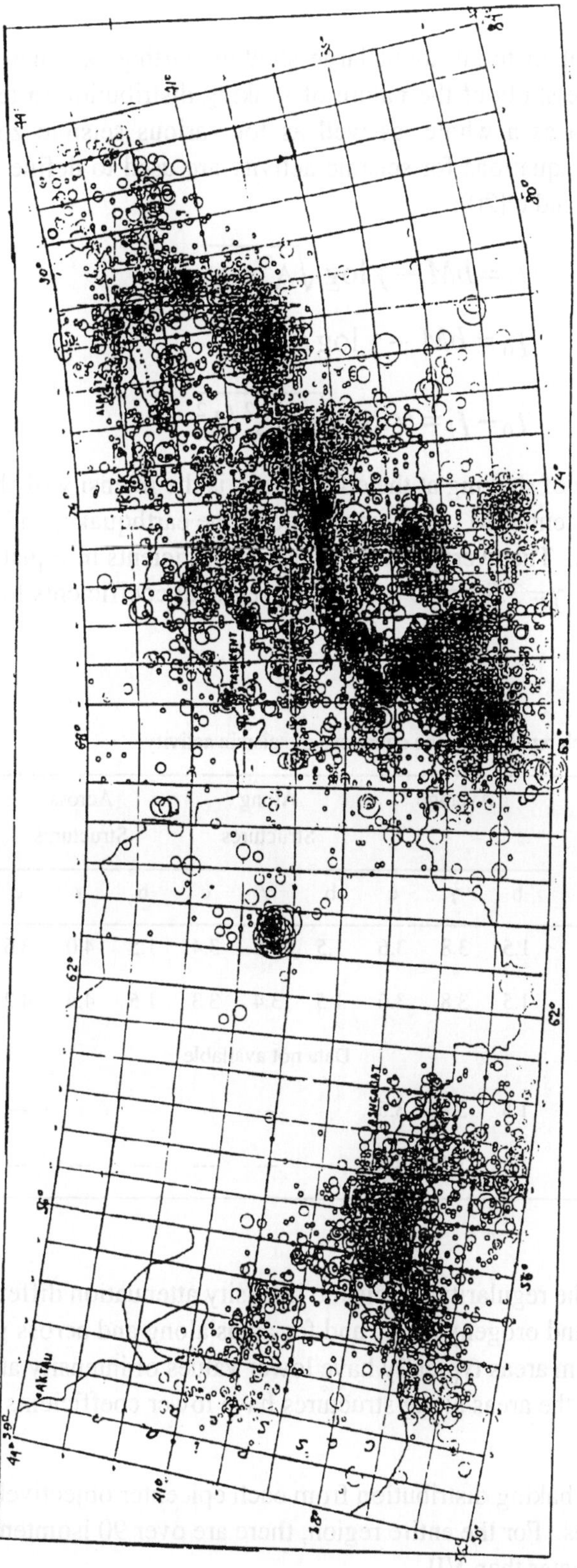

Figure 5. Map of epicenters of weak shallow earthquakes ($M \geq 3.0$; $H \leq 70$ km).

4. Seismic Intensity Attenuation in Central Asia

A significant amount of material about large shallow earthquakes has been gathered. These data facilitate the study of the nature of shaking distribution in terms of seismic intensity for the region as a whole, as well as for various seismic zones within the region. The following equations for seismic activity are used to define regional values of the coefficients γ, b, and c [20]:

$$I_i = bM - \gamma \log \sqrt{\Delta^2 + h^2} + c \tag{1}$$

$$I_0 = bM - \gamma \log h + c \tag{2}$$

$$I_0 - I_i = \gamma \log \sqrt{1 + \Delta^2 / h^2} \tag{3}$$

where I_o is the intensity of shaking at the epicenter, I_i is the intensity of shaking at the distance Δ from the epicenter, M is the magnitude of the earthquake, and h is the depth of the earthquake focus. The average values for the coefficients in Equations 1, 2, and 3 are reported as: b = 1.5, γ = 3.5, and c = 3.0. The regional coefficients for each republic are shown in Table 2.

TABLE 2. Parameters in each republic for the equations of seismic activity

	Average			Along Structures			Across Structures		
Republic	b	γ	c	b	γ	c	b	γ	c
Kazakhstan	1.5	3.8	3.6	1.5	3.6	3.4	1.5	4.0	3.6
Kirgizstan	1.5	3.8	3.6	1.5	3.4	3.3	1.5	4.4	4.2
Tajikistan				Data not available					
Turkmenistan	1.5	3.8	3.5	---	---	---	---	---	---
Uzbekistan	1.5	3.1	2.5	---	---	---	---	---	---

It was found that the regularity of seismic intensity attenuation differs significantly between the platform and orogeny areas, and for areas along and across geologic structures. The platform areas typically have lower values of intensity attenuation than the orogeny areas, and the areas along structures have lower coefficients than those across structures.

The nature of the shaking distribution from each epicenter objectively reflects the individual isointensities. For the entire region, there are over 90 isointensity maps for earthquakes with I_o greater than VII.

For the purpose of seismic hazard assessment of the whole region, it is of interest to define a summary of seismic activity, that is the areas that have been affected by various intensities. Such an assessment can be made with the help of a map of combined isointensities as shown in Figure 6, created from maps of several large earthquakes. Isointensity lines are primarily oriented in the direction of the tectonic structures in the region. The largest areas of maximum shaking with intensity I_o greater than IX are due to the well known Krasnovodsk, Yerevan, Chilik, Kemin, and Kashgar earthquakes.

It is not yet possible to report about similarities of attenuation, expressed in terms of the parameters of strong ground motion recordings, because there are very few such recordings for large earthquakes in Central Asia. The ability to record strong ground motion has developed only in the past 20 years. By 1990, there were about 200 strong ground motion recording stations in the entire region [22]. Roughly 800 recordings have been made from approximately 150 earthquakes; however, most of the recordings were taken far from the source of the earthquake. The strongest recordings are those of the M 7.3 Gazli earthquake on May 17, 1976, which were taken in Karakir, Uzbekistan near the epicenter zone. Peak ground accelerations for this recording are 0.8g in the horizontal direction and 1.3g in the vertical direction, with a shaking intensity of X. Another event, the M 6.3 Baisorun earthquake on November 12, 1990 was recorded at Kurmenti, located 30 km from the epicenter. The peak horizontal acceleration for this recording was 0.65g, with a seismic intensity of VIII.

In order to estimate the expected strong ground motion parameters for regions in Central Asia, extrapolations are made from relationships developed for other parts of the world. For example, relationships developed by Aptikayev [23], Fukushima, and Tanaki [24] are often used.

5. Methods for Seismic Hazard Assessment in Central Asia

The concept of seismic hazard assessment in Central Asia has been in place since the 1960's. It is based on the conventional principle of modeling the occurrence of earthquakes at the source and the shaking on the surface of the earth. Work on seismic zoning includes a two-stage approach [25,26]. The first stage includes the definition of real and potential zones for the generation of seismic events. In the second stage, the expected parameters of ground shaking on the surface of the earth and the probabilities of their occurrence are estimated. Seismic hazard is assessed in terms of the distribution of probable seismic actions (in units of seismic intensity and parameters of ground motion) according to their distribution in time and space.

An important component in seismic hazard assessment in Central Asia is seismic zoning, which is divided into three types [27]: general seismic zoning (GSZ), detailed seismic zoning (DSZ), and seismic micro zoning (SMZ). The differences between the types are in the content of the tasks, methods for carrying out the tasks, and objects for investigation. These differences lead to different scales of mapping.

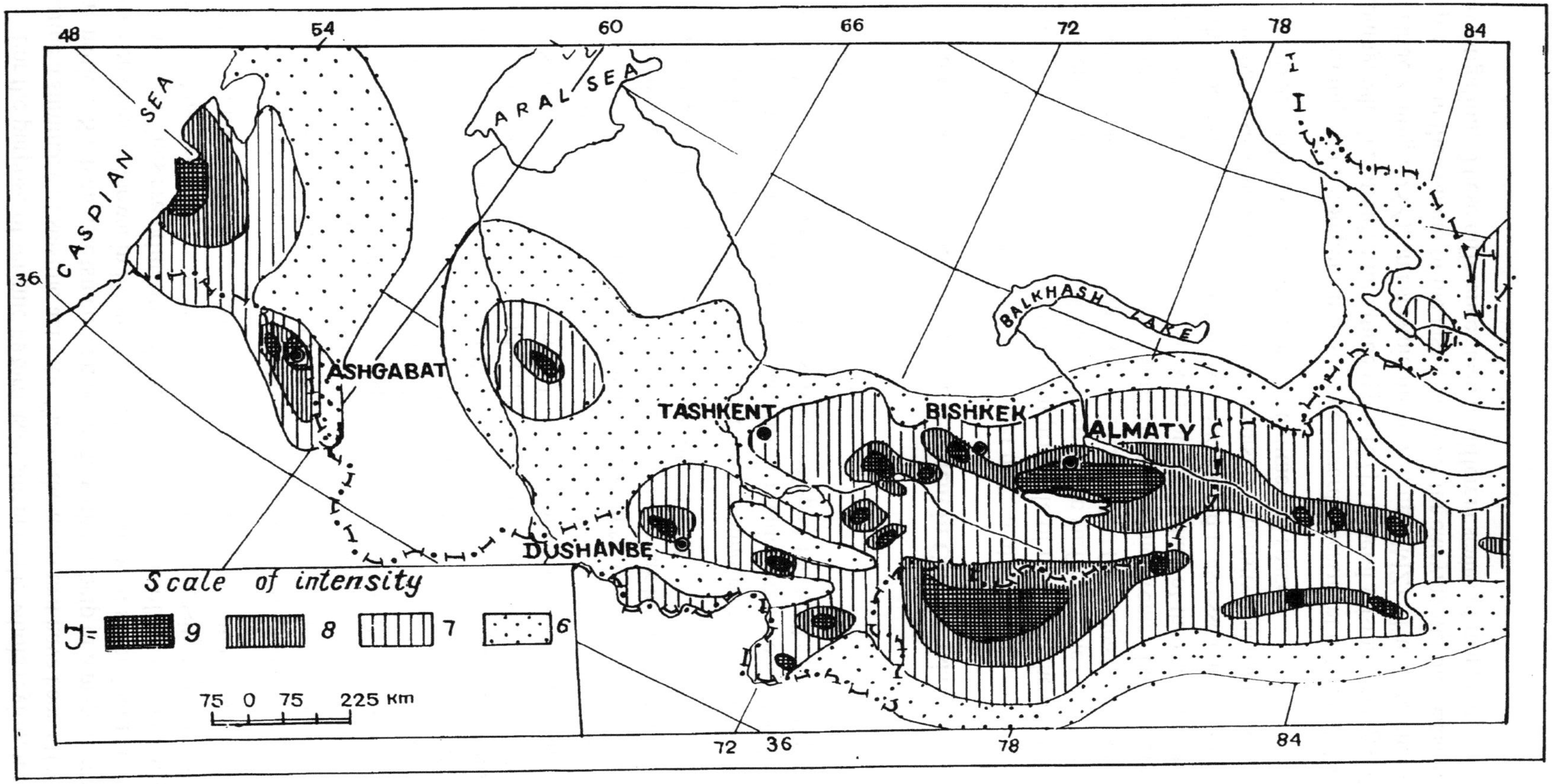

Figure 6. Summary isoseismal map of strong shallow earthquakes.

General seismic zoning (GSZ) is intended for studying large seismic generating structures that define the seismicity of the whole regìon. Using the seismic hazard assessments shown in a GSZ map, it is possible to forecast damages to the structures of mass construction. It is assumed that significant losses due to damages of this type are caused by earthquakes with magnitudes greater than 6.1.

Detailed seismic zoning (DSZ) is intended for studying seismic generating structures that present hazard for a specific location. A specific location can include an individual structure, a group of structures, a settlement, or a region of prospective economic development. With DSZ, the possibility of smaller earthquakes occurring near the study site is investigated, because the seismic intensity can be high in a small region for earthquakes of relatively small magnitude.

The intent of seismic micro zoning (SMZ) is to consider the influence of local conditions on the seismic action. Local conditions include topography, composition and structure of environment, presence of groundwater, and other factors that may influence the ground shaking at the surface. In SMZ, special attention is paid to the forecasting of geological hazards associated with earthquakes, such as liquefaction, landslide, subsidence, and rockfall.

GSZ, DSZ, and SMZ are integral parts of the work of seismic hazard assessment. The results are presented in the form of maps with isolines of seismic action, often including the probability of occurrence of the action.

For more than 50 years, seismic hazard zoning maps, which are the most important result of the fundamental and applied seismic research, are included in the documents used to regulate design and construction in seismically active regions. They are updated and improved as new information on earthquakes and seismic hazard becomes available. Figure 7 shows part of the GSZ map developed in 1978 for the USSR territory (GSZ-78). The map includes:

- Boundaries of shaking intensity with zones for I = VI, VII, VIII, and IX
- Zones of the most probable locations of severe earthquakes, differentiated by the maximum expected magnitude, ranging from 6.1 to greater than 8.1
- Data on probability of shaking in zones with equal seismic data; for example, 8_1, 8_2, and 8_3 refer to a magnitude 8 event once every 100, 1000, and 10000 years, respectively

As shown in Figure 7, the regions of North Tien-Shan and Krasnovodsk fall in the zones of largest seismic potential. The contours of these zones, as well as all other zones, have been established according to the available geological, geophysical, and seismological data.

It should be emphasized that all of the capital cities of the Central Asian republics are located in regions of high seismic potential. Almaty, the capital of Kazakhstan, Ashgabad, the capital of Turkmenistan, Dushanbe, the capital of Tajikistan, and Bishkek, the capital of Kyrgyzstan, are all located in zones with potential seismic

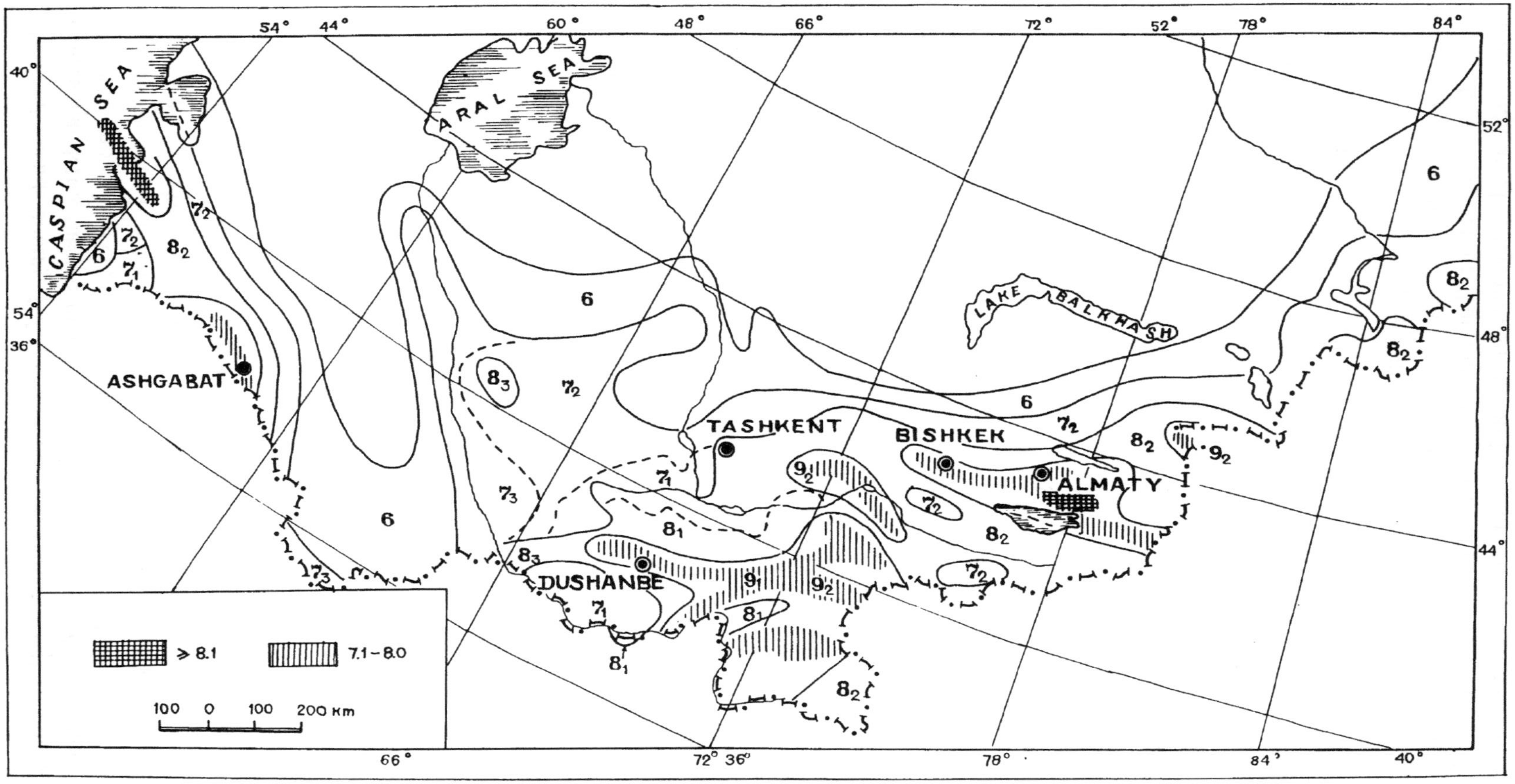

Figure 7. Map of seismic zoning of the Central Asian region.

intensity of IX. Tashkent, the capital of Uzbekistan, is located in the zone with potential seismic intensity of VIII.

The intensities associated with the GSZ and DSZ maps are for average ground conditions (soils of the 2nd category according to the Classification of Construction Norms and Regulations). For estimating intensities for other ground conditions, it is necessary to use information in accordance with SMZ. For the majority of the cities and large settlements, seismic hazard microzoning that takes into account the influence of local soil conditions has been done.

In order to use the seismic hazard information of zoning maps for assessing building vulnerability, the design and construction regulations include a translation from seismic intensity (I) to maximum horizontal acceleration (A). The translation is based on observed data as follows: I = IX corresponds to A = 400 cm/s^2; I = VIII corresponds to A = 200 cm/s^2; and I = VII corresponds to A = 100 cm/s^2. An analysis of the large volume of worldwide seismic data shows that the translation should assign a range of maximum horizontal acceleration values to each intensity level. The most probable value for maximum horizontal acceleration (A) at each intensity level (I) is found to be: I = IX corresponds to A = 580 cm/s^2; I = VIII corresponds to A = 300 cm/s^2; and I = VII corresponds to A = 140 cm/s^2 [30].

There are still many unsolved problems with the methods and practice of seismic hazard assessment in the Central Asia region. Even in areas with well-known geological, geotechnical, and seismological information, the mapped seismic hazard zones have been underestimated in some instances. Large earthquakes have occurred in mapped zones of relatively low seismic potential. This has occurred in other regions of the former USSR, for example Armenia in 1988, Georgia in 1991, and Sakhalin in 1995. Incorrect seismic hazard assessment, coupled with poor construction quality, lead to heavy damage and loss.

Since 1991, work has been underway on the development of a new seismic hazard map for territories of the former USSR (GSZ-98). The map should meet current international standards and include recent information on regional seismicity. It is expected that this new map will be more informative than all previous maps and allow for more reliable assessments of seismic hazard.

Seismic hazard zoning is a problem for territories with low seismicity where there is a great lack of geological and seismological information. In addition, problems also exist in regions of intensive exploration of gas and oil deposits, where seismic hazard may be increased by technological factors. These regions exist in Kazakhstan, Uzbekistan, and Turkmenistan.

6. General Data on Capital Cities of Central Asian Countries

The seismic hazard assessment of the capital cities of the Central Asian republics (i.e., Almaty, Bishkek, Tashkent, Dushanbe, and Ashgabad) should be done in a very detailed manner. In general, these are the largest cities, with more than 5 million inhabitants and a large number of industrial facilities and educational, cultural, and scientific institutions. In the capital cities, more than 70% of the residential buildings are multistory, and many have not been designed to current seismic requirements.

Table 3 presents the general data on the capital cities of the Central Asian republics. The information includes the date the city was founded, its geographical coordinates, size, and number of inhabitants. The date the city was founded is taken as the year when the city was officially established, even though a settlement may have previously existed there for several years.

TABLE 3. General data on capital cities of Central Asian republics

			Coordinates		Size (km^2)		Population (x 1,000,000)	
Republic	Capital City	Date Founded	Lat (°N)	Long (°E)	Total	Multi-story buildings	Total	Multi-story buildings
Kazakhstan	Almaty	1867	43.23	76.95	300	50	1.5	1.0
Kyrgyzstan	Bishkek	1825	42.82	74.63	173.2	---	0.783	0.4
Tajikistan	Dushanbe	1924	38.57	68.80	140	---	1.1	---
Turkmenistan	Ashgabad	1881	38.00	58.20	170	68	0.5	0.3
Uzbekistan	Tashkent	12th cent.	41.33	69.25	256	120	2.2	1.0

7. Large Earthquakes Affecting the Capitals Cities of Central Asian Republics

As shown in Figure 7, all the capital cities of the Central Asian republics are located in zones of high seismicity. In the past, these cities have been affected by strong earthquakes of intensities VIII, IX, and in some cases X, which caused heavy damage and loss. Almaty, Kazakhstan twice experienced shaking of intensity X, in 1887 [21], which devastated the city, and in 1911 [22]. Ashgabad, Turkmenistan was totally destroyed in 1948, when shaking of intensity IX led to the deaths of 40,000 inhabitants.

Table 4 shows a catalog of earthquakes of intensity VII or greater that have affected the capital cities of the Central Asian republics. Figures 8a through 8g show maps of the epicenter location and intensity distribution for several of the earthquakes listed in Table 4. The distribution of intensity helps to illustrate the spatial orientation of the earthquake mechanism and to assess some of the seismic data such as magnitude, depth, and location.

TABLE 4. Main parameters of large earthquakes that have affected capital cities of Central Asian republics

Capital City	Date	Earthquake Name	Coordinates		Magnitude	Intensity in Capital City	Distance to Epicenter (km)	Number of lives lost	% of buildings damaged
			N. Lat (°)	E. Long (°)					
Almaty	1807	Almaty	43.1	76.9	6.7	VIII-IX	15	---	---
	6/8/1887	Verny	43.1	76.8	7.3	IX-X	13	342	100
	7/11/1889	Chilic	43.4	78.4	8.3	VII-VIII	100	---	27
	1/3/1911	Kemin	42.9	78.9	8.2	IX-X	50	---	40
Bishkek	8/2/1885	Belovodsk	42.4	74.1	6.9	VII-VIII	45	---	95
	7/9/1889	Pishpek	42.8	74.9	5.8	VI-VII	25	---	---
Dushanbe	2/27/1952	Stalinabad	38.8	68.9	4.7	VI-VII	12	---	---
Ashgabad	2000 BC	Ak-Tepi	38.0	58.2	7.1	IX	18	---	---
	10 AD	Nisskoye-1	38.0	58.3	7.1	IX	15	---	---
	953	Nisskoye-2	38.0	58.3	7.1	IX	15	5000	---
	11/17/1893	Kurgan-1	37.2	58.4	6.6	VII	100	18000	---
	5/1/1929	Gifan/Germab	37.8	57.7	7.2	VIII	80	3000	---
	10/5/1948	Ashgabad	37.9	58.3	7.3	IX	20	40000	100
Tashkent	2/4/1868	---	41.0	69.0	5.6	VII	40	---	---
	4/3/1868	Tashkent-1	41.2	69.5	6.7	VII-VIII	27	---	---
	11/29/1886	Tashkent-2	41.4	69.5	6.7	VII	30	---	---
	11/28/1888	Kostakoz	40.0	69.6	6.3	VI-VII	145	---	---
	11/17/1897	Ura-Tubin	39.8	68.4	6.6	VI-VII	170	---	---
	6/7/1924	---	41.3	69.2	4.3	VII	9	---	---
	4/25/1966	Tashkent-3	41.3	69.3	5.1	VIII	0	---	60
	12/11/1980	Nazarbek	41.3	69.1	4.8	VI-VII	16	---	---

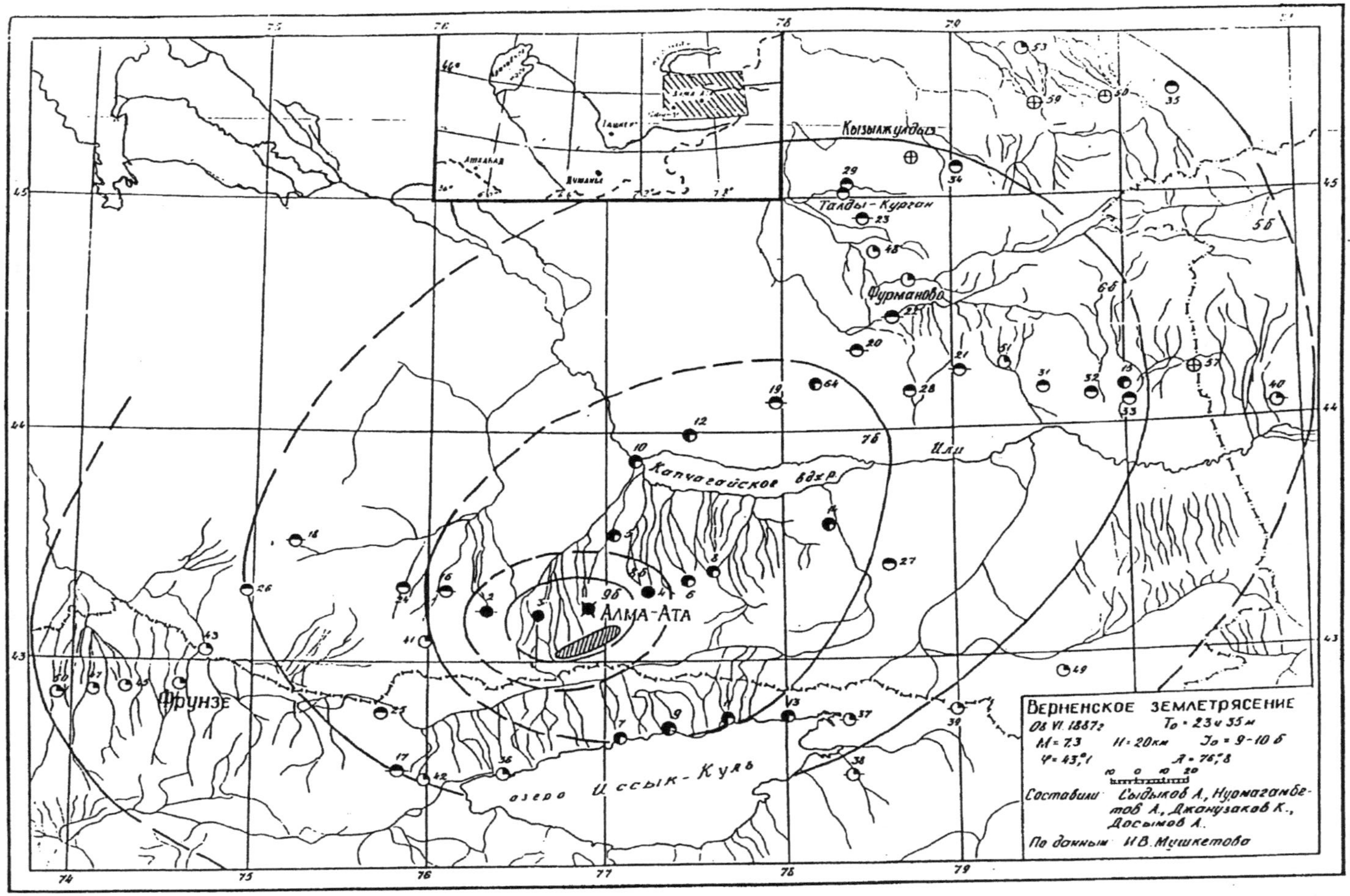

Figure 8a. Isoseismal map of Vervy earthquake of 1887.

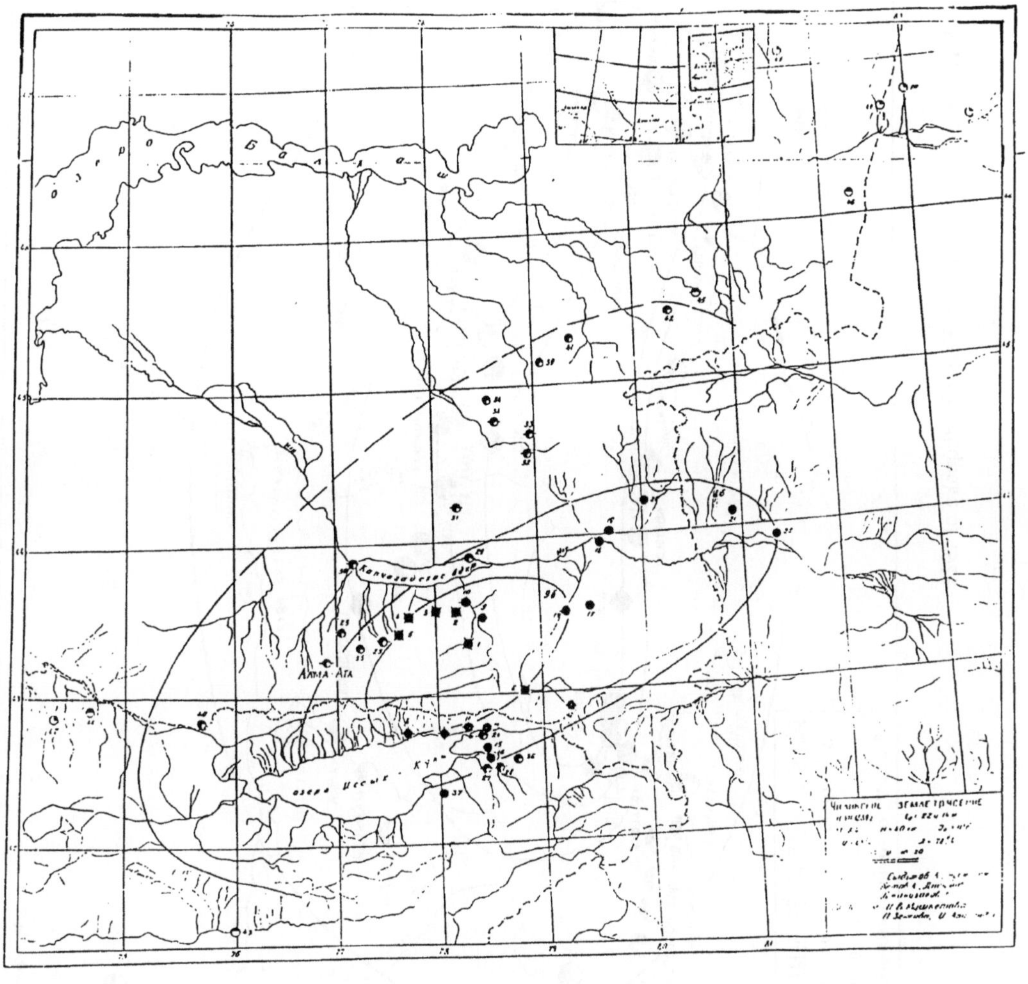

Figure 8b. Isoseismal map of Chilik earthquake of 1889.

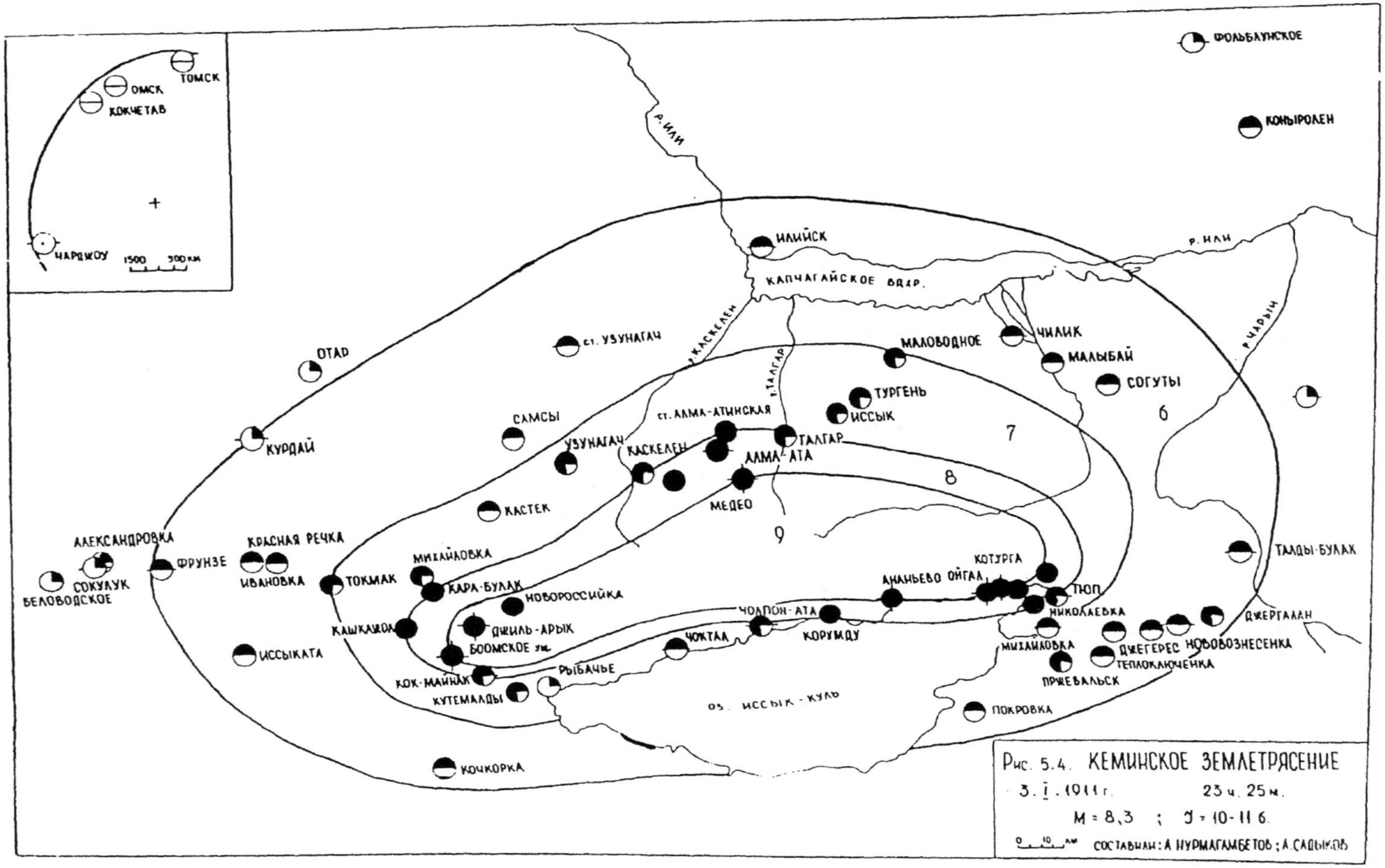

Figure 8c. Isoseismal map of Kemin earthquake of 1911.

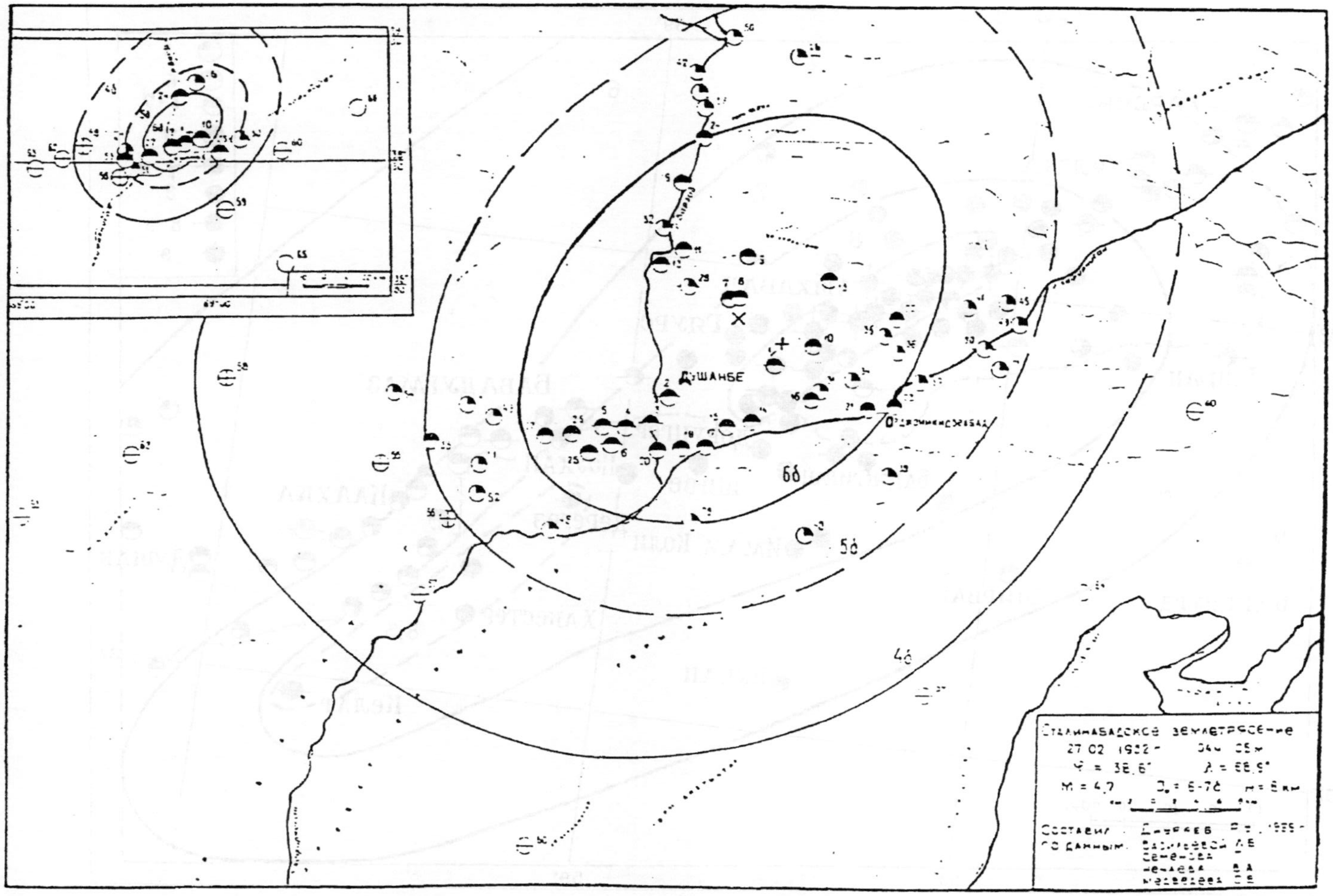

Figure 8d. Isoseismal map of Stalinabad earthquake of 1952 (R. Djuraev).

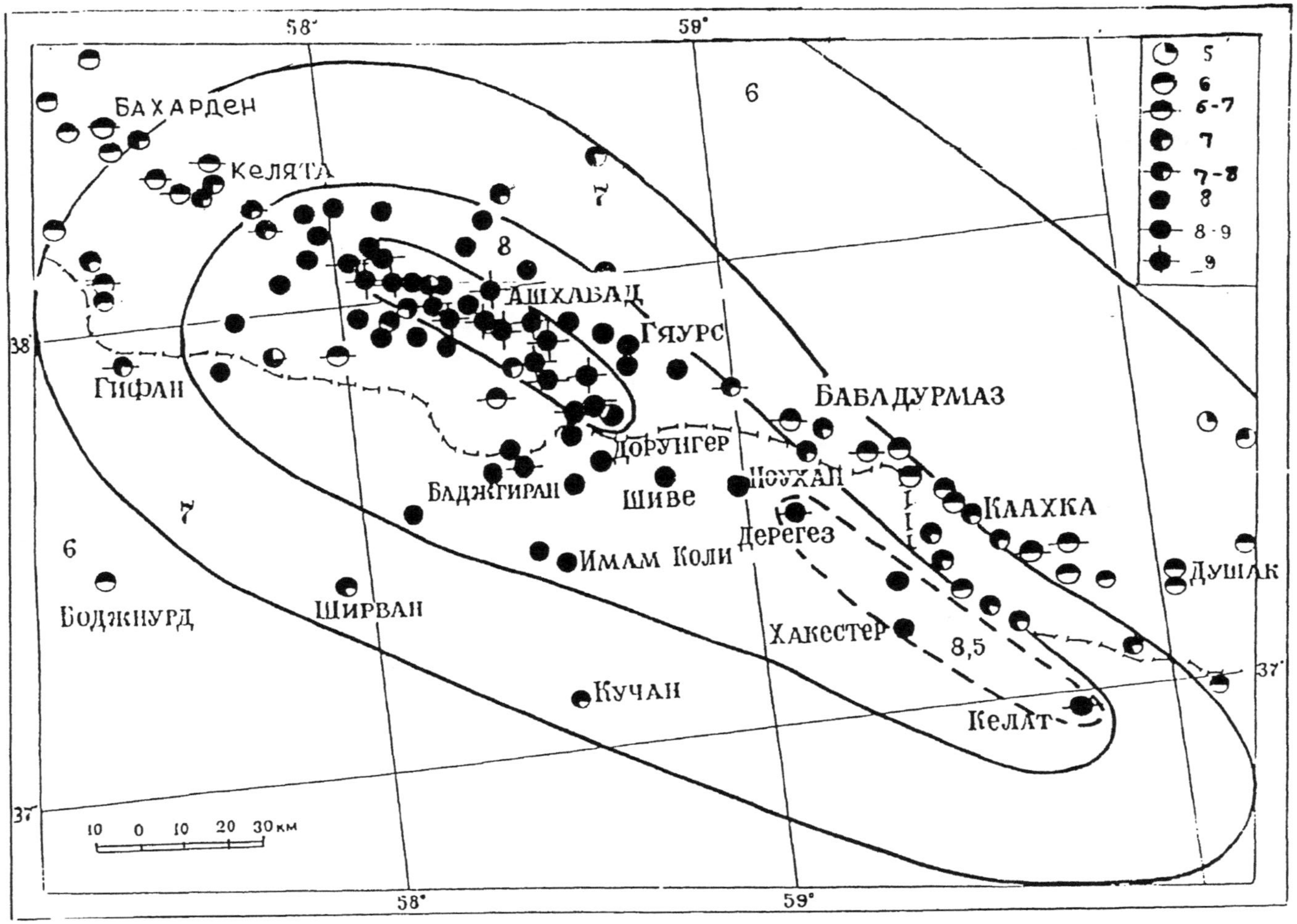

Figure 8e. Isoseismal map of Ashgabad earthquake of 1948 (G. Golinsky).

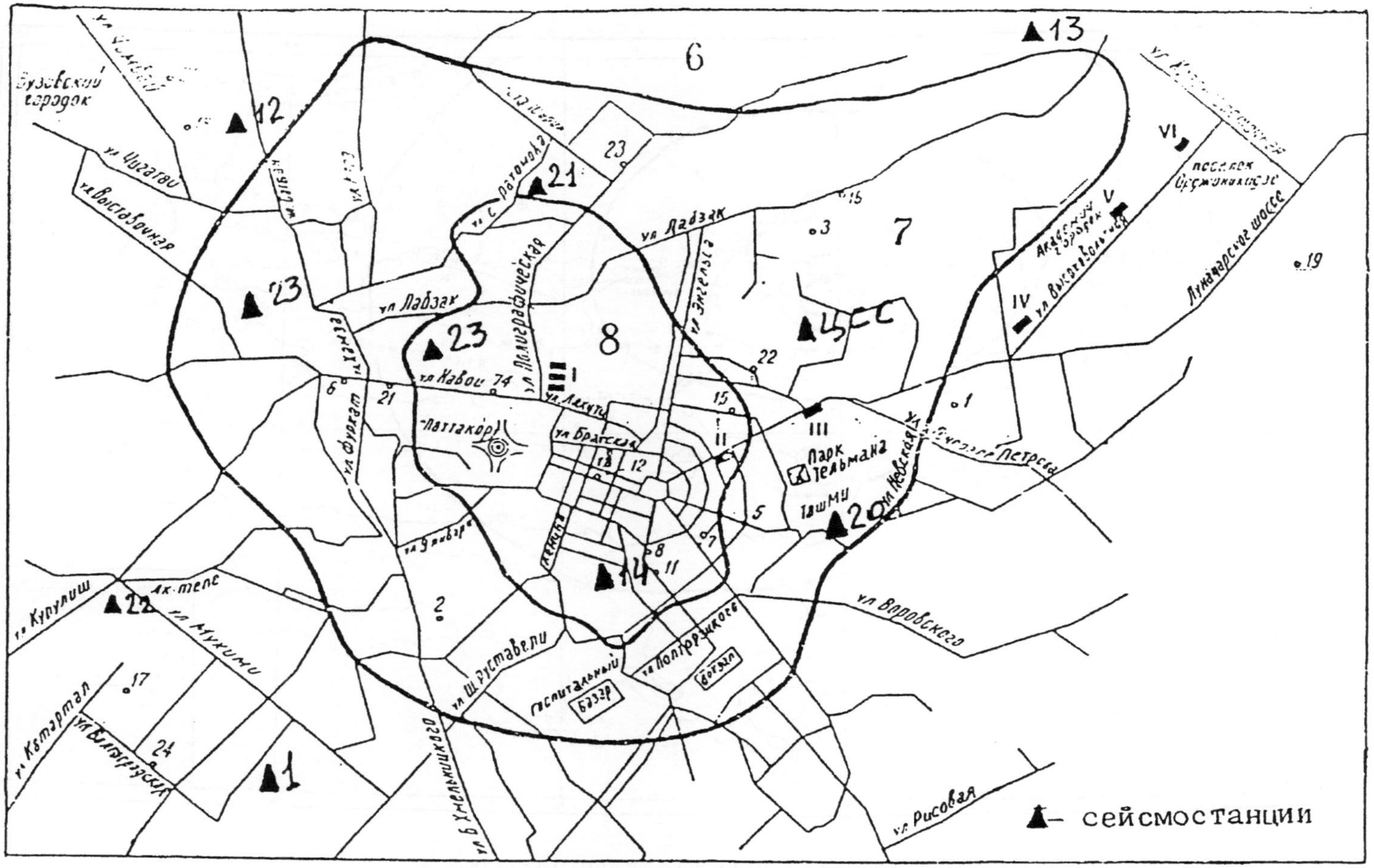

Figure 8f. Isoseismal map of Tashkent earthquake of 1966 (V. Rasskazovsky, T. Rashidov, K. Abdurashidov).

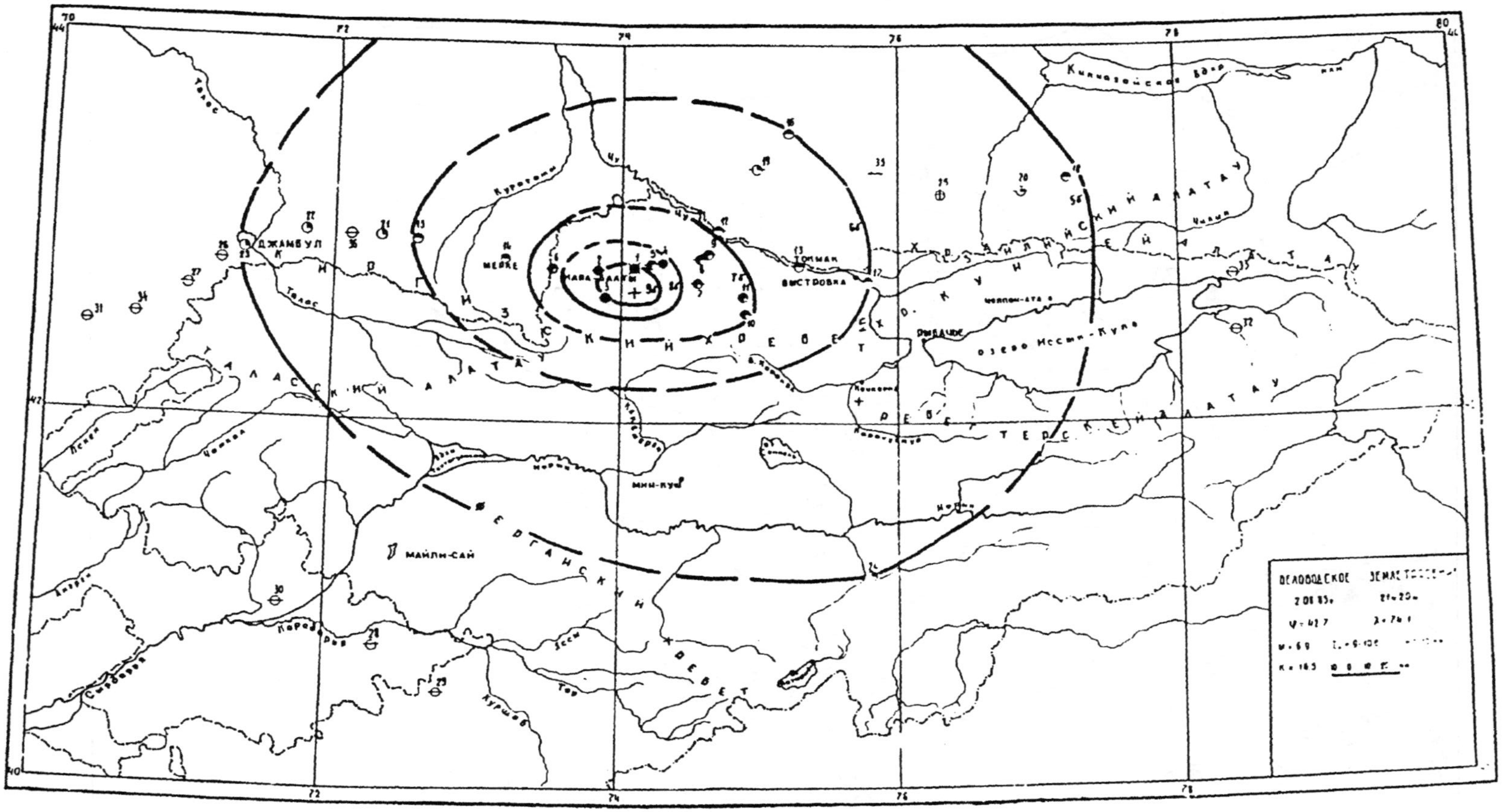

Figure 8g. Isoseismal map of Belovodsk earthquake of 1885.

8. Information on Seismic Source Zones in Central Asian Republics

Research has been done in the capital cities of the Central Asia republics to identify the seismic generating zones. The locations and activities of active seismic zones have been identified [27]. Information such as maximum possible earthquake magnitude, type of earthquake mechanism, and potential depth of epicenters has been defined for most of the seismic source zones.

For the capital cities of Almaty and Bishkek, detailed seismic zoning of the city and surrounding territories has been done [31, 32]. Microzone maps showing the primary seismic source zones and distribution of intensity and acceleration have been made and are shown in Figures 9a and 9b. For the capital cities of Ashgabad, Tashkent, and Dushanbe, detailed seismic zoning has not been done. Seismic source zones in Ashgabad were defined according to observed earthquake events (see map in Figure 9c). In Tashkent and Dushanbe, seismic source zones were defined according to the presence of faults and tectonic movements (see Figures 9d and 9e).

Table 5 lists summary information about the seismo-tectonic situation in the regions of the capital cities. Table 5 includes names of seismic source zones, maximum expected magnitude for each zone, and distance of the zone to the capital city. This information helps to characterize the seismic hazard that exists in each capital city. Almaty is the only city that is shown to have the possibility of experiencing nearby earthquakes of magnitude greater than 8.0. Earthquakes of magnitude 7.5 are shown to be possible near the capital cities of Ashgabad, Bishkek, and Dushanbe, while the maximum earthquake magnitude that might affect Tashkent is assumed to be 6.8. Based on the information in Table 5, the maximum earthquake shaking intensity that could be experienced in Almaty, Bishkek, Dushanbe, and Ashgabad is IX, and in Tashkent is VIII.

9. Seismic Microzoning in Capital Cities of Central Asian Republics

All of the capital cities have had some type of seismic micro zoning studies completed for them. The local geological conditions of the different territories were studied to identify the soil conditions, topographic characteristics, location of ground water, and potential for effects such as liquefaction, subsidence, and landslide. The zoning of the seismic hazard maps includes the effects of shaking as well as the increase due to potential hazards caused by local soil conditions.

Figure 10a shows the seismic hazard map for the city of Almaty, indicating areas with potential for seismic intensities of VIII, IX, and X [33]. The map is based on the observations of shaking due to earthquakes in 1887, 1889, and 1911, seismic recordings at the local stations, and engineering geology information. In addition, the city contains several faults that have not been associated with any previous earthquake but have the potential to generate large events.

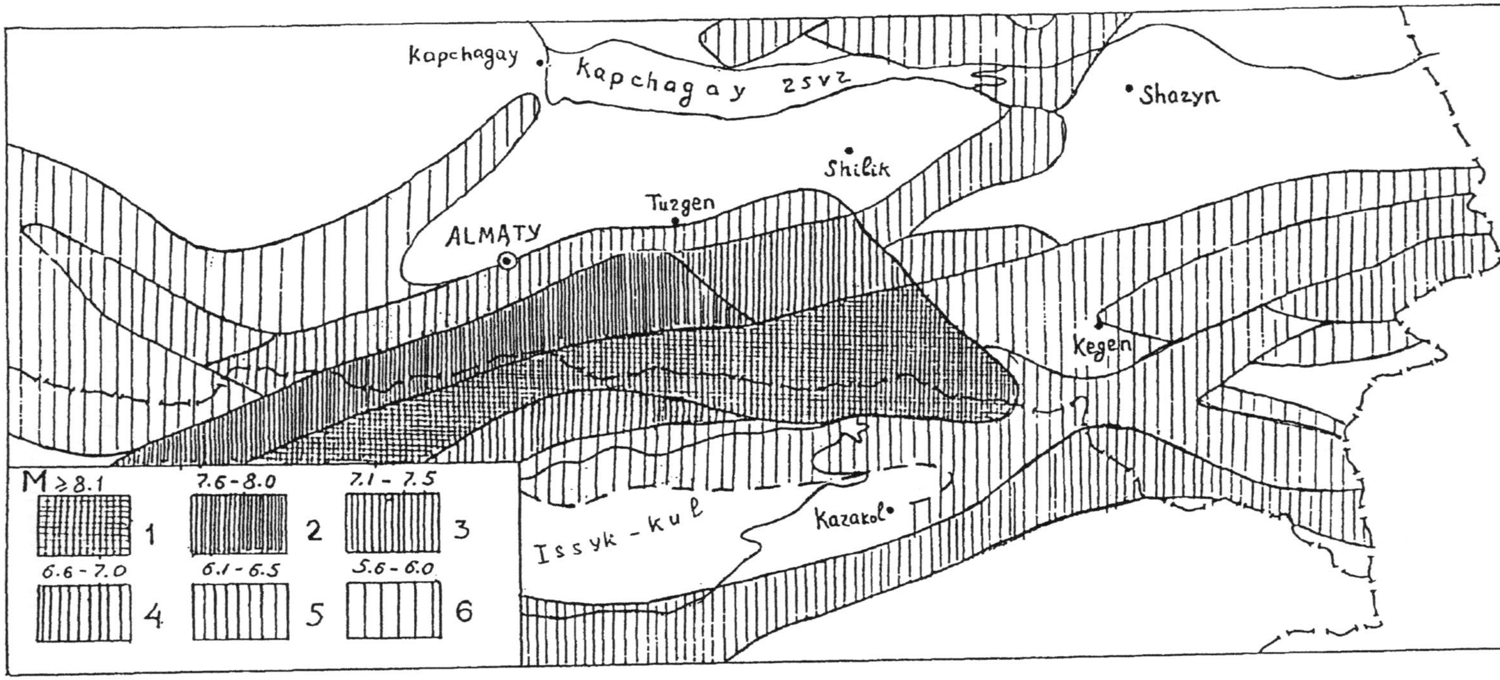

Figure 9a. Map of seismic generating zones of Almaty industrial region.

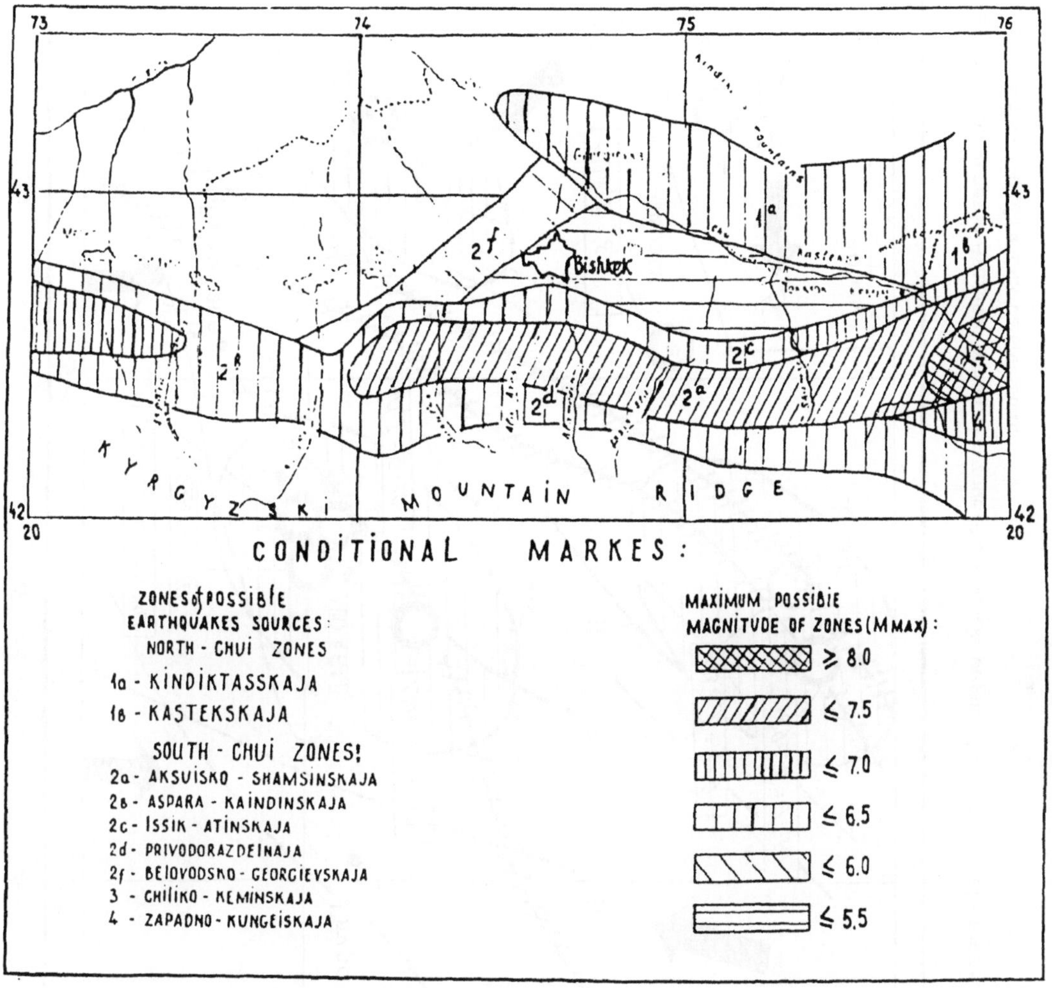

Figure 9b. Map of seismic generating zones of Chu depression and the mountain environment.

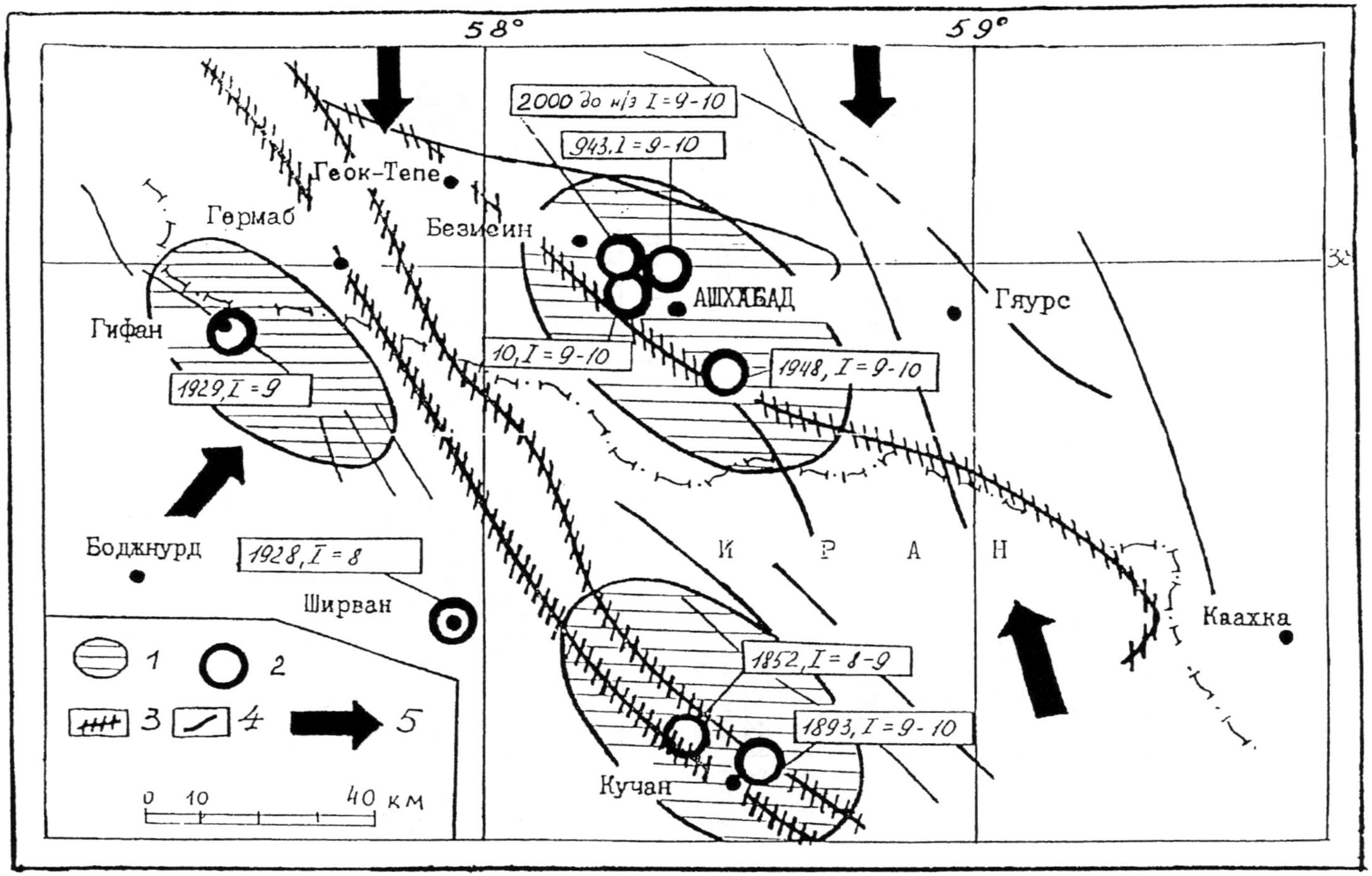

Figure 9c. Map of possible seismic zones in Ashgabad region. 1 = zone of earthquake origins; 2 = epicenters of strong earthquakes; 3 = main deep fault zones; 4 = local deep fault zones; and 5 = direction of tectonic compression.

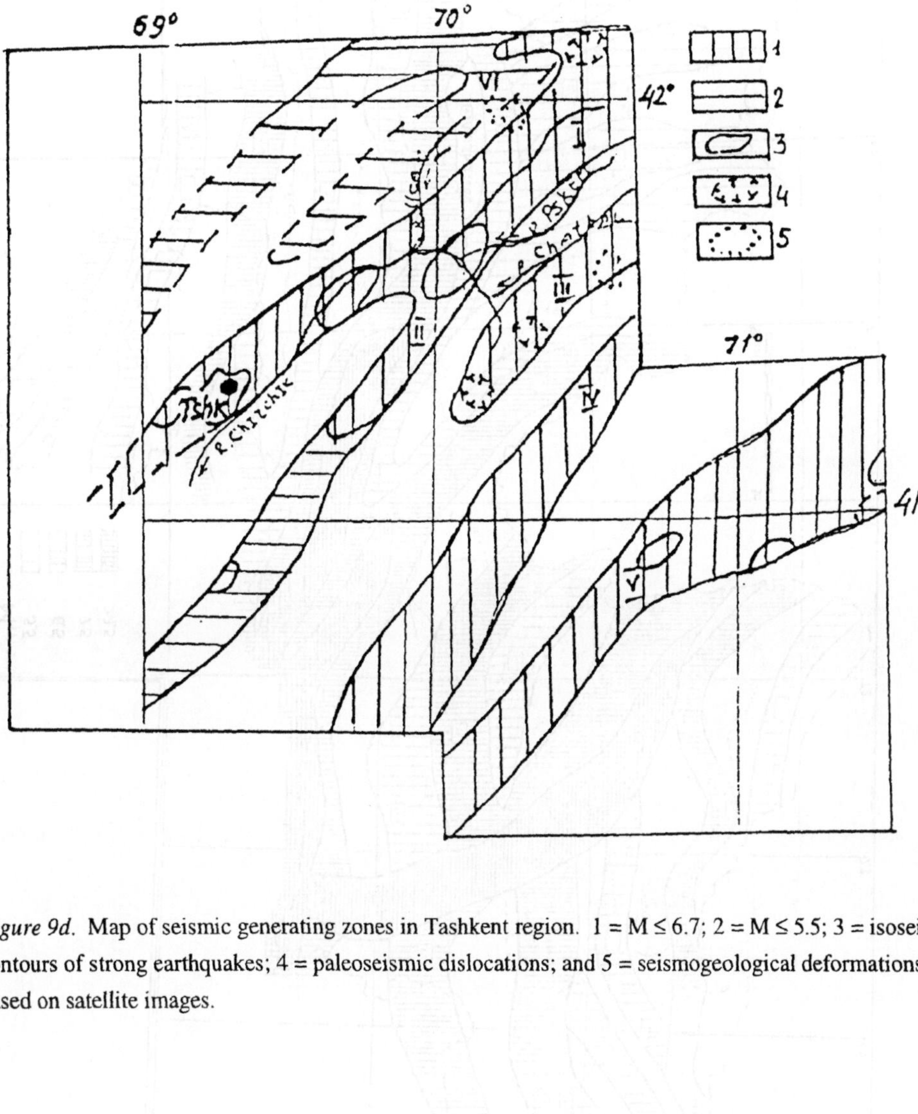

Figure 9d. Map of seismic generating zones in Tashkent region. 1 = M ≤ 6.7; 2 = M ≤ 5.5; 3 = isoseismal contours of strong earthquakes; 4 = paleoseismic dislocations; and 5 = seismogeological deformations based on satellite images.

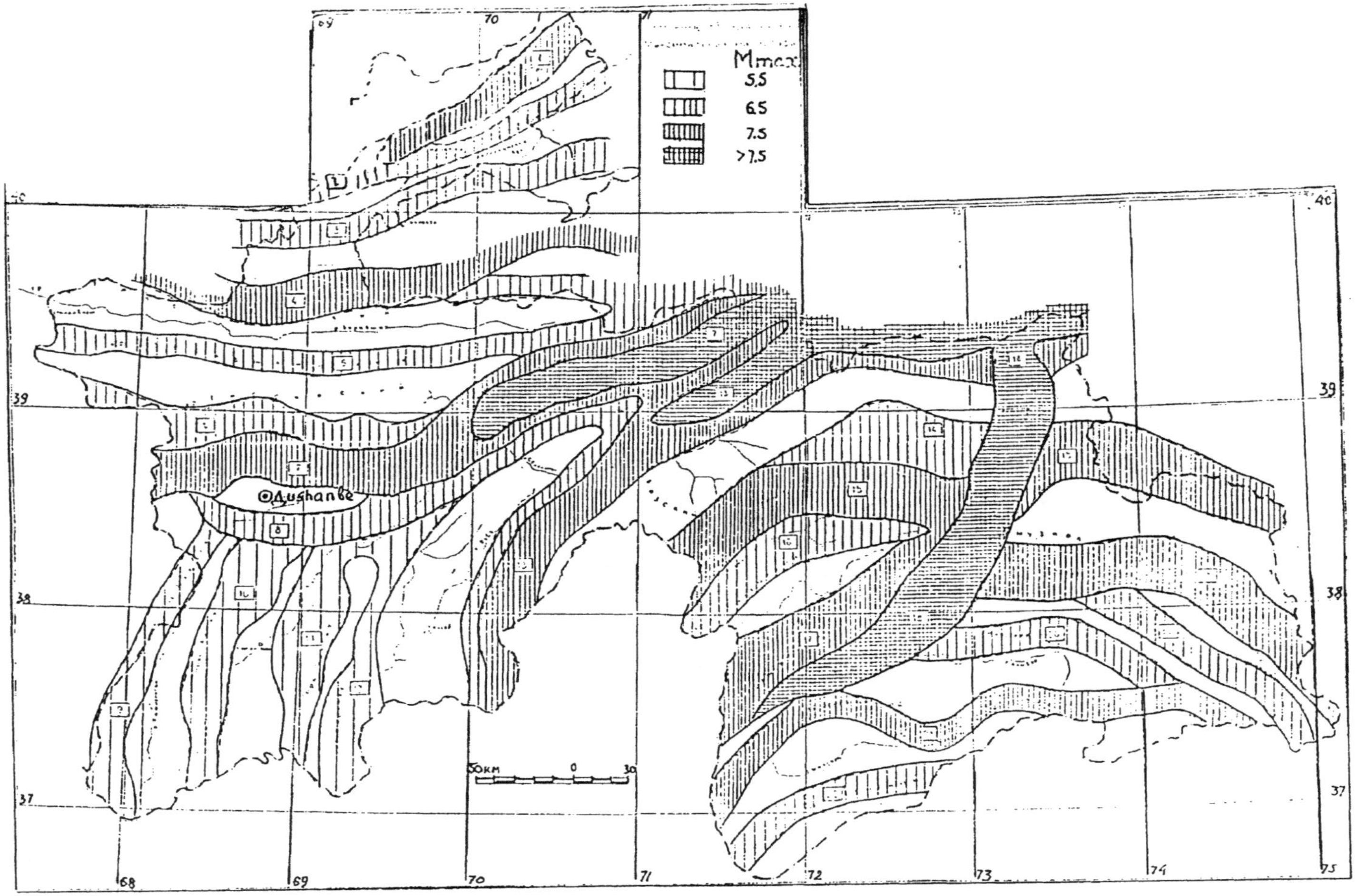

Figure 9e. Map of seismic generation zones of Tajikistan.

TABLE 5. Data on seismic source zones in capital cities of Central Asian republics

City	Zone Name	Maximum Magnitude	Recurrence Interval (years)	Distance to City (km)	Earthquake Mechanism
Almaty	Kungey	8.2	500-725	42	Thrust faulting / strike slip
	Zailisk	8.0	600-680	25	Thrust faulting / strike slip
	Almaty	7.0	3700	7	Thrust faulting / strike slip
Bishkek	Kindiktass	6.5	30	26	Thrust faulting
	Kastek	7.0	100	80	Thrust faulting
	Aksu-Shamsi	7.5	300-600	24	Thrust / normal faulting
	Issik-Aty	6.5	30	21	Thrust faulting
	Privodorazdel	6.5	30	34	Thrust / normal faulting
	Belovidsk-Georgiyevka	6.0	30	23	Thrust / normal faulting
	Chilik-Kemin	8.0	1500-2000	100	Thrust faulting
Dushanbe	South-Gissar	7.5	---	3.5	---
	Ilek-Vaksh	6.5	---	20-25	---
Ashgabad	Ashgabad	7.5	300	8	Normal faulting / strike-slip
	Germab	7.2	300	50	Normal faulting / strike-slip
	Kurgan	7.0	60	55	Normal faulting / strike-slip
Tashkent	Karzhantau	6.8	---	0-5	Thrust faulting / upthrust
	Nurekaty	6.4	---	20-75	Thrust faulting / strike-slip
	Sandalash-Chetkal	7.5	---	160-270	Thrust faulting

The central part of Almaty is composed of rock covered with sandy and loamy deposits. The Paleozoic foundation is located at a depth of about 2500 meters, and groundwater is located at depths ranging from 15 to 120 meters. All of this territory has the potential for seismic intensity IX. In the north part of the city, the soil includes thick layers of sand and loam. Below the depth of 20 meters, there are sandy deposits that include crushed rock. Groundwater in this area is located at depths from 3 meters to more than 10 meters. The intensity of shaking in this region is estimated at IX to X. In the northwest part of Almaty, the intensity is reduced to VIII because of the distance to the primary seismic generating zones of Zailisk and Kungei Alatau, which are located to the south of the city.

Figure 10b shows the seismic hazard map for the city of Ashgabad [34]. Intensities in the city range from VIII to more than IX. The south and west parts of the city have rather favorable soil conditions with rock overlaid by sandy deposits. The depth of groundwater is about 15 meters, but has been rising in the past few years. The north

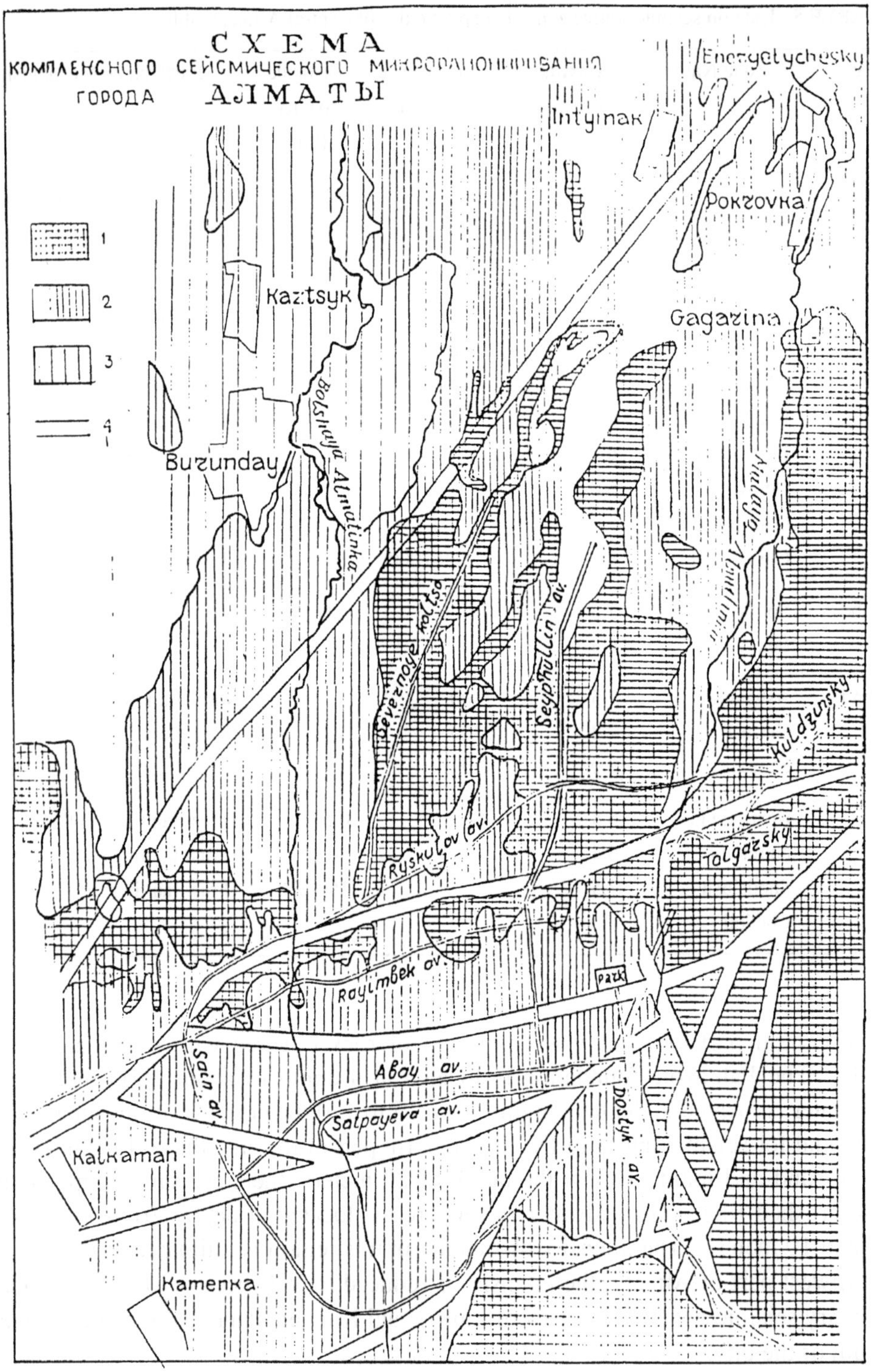

Figure 10a. Map of seismic microzoning in Almaty. 1 = intensity 10; 2 = intensity 9; 3 = intensity 8; and 4 = zones of possible surface faulting [33].

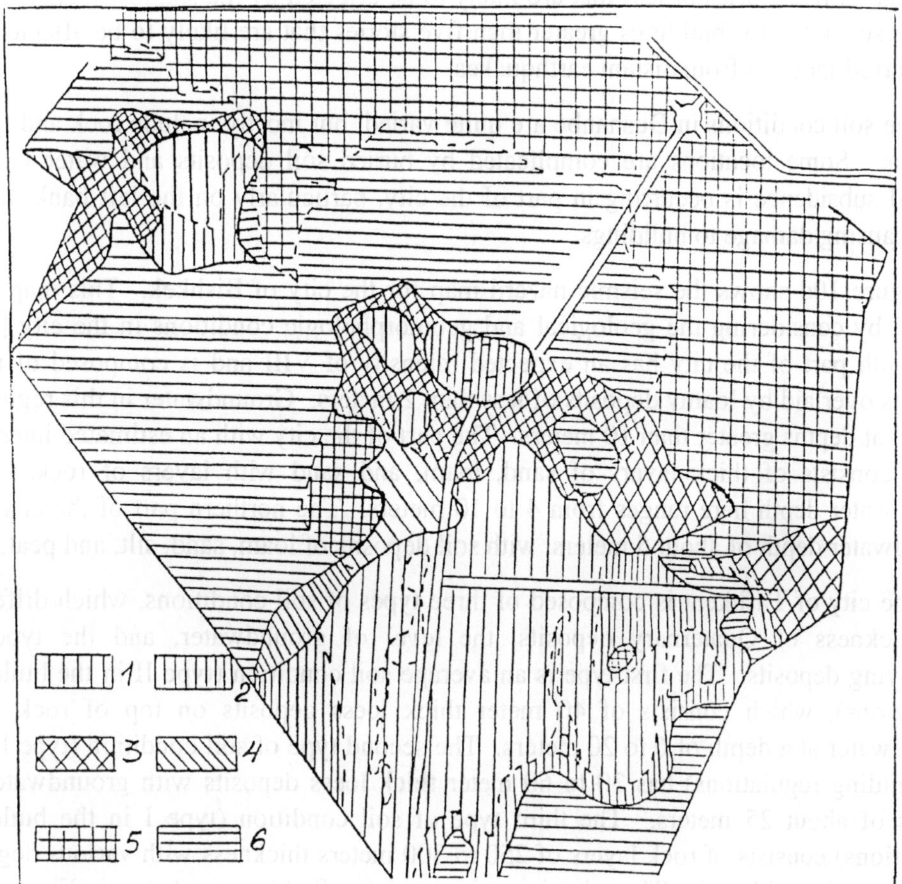

Figure 10b. Map of seismic microzoning in Ashgabad territory [34]. 1, 3, 5 = zones of intensity 8, 9, and more than 9; 2, 4, 6 = regions with complicating factors inside the zones.

and east parts of the city have unfavorable soil conditions and a high water table located at a depth of about 3 meters. For this reason, the intensity in this area is higher than in the south and west parts of the city.

The city of Dushanbe has seismic hazard from shallow local earthquakes as well as deep focus distant earthquakes. These two types of potential seismic events will produce ground motions with differing spectral content. For this reason, two different seismic hazard zoning maps have been made as shown in Figure 10c. The first is for buildings of five stories or less that are likely to be affected by the shallow local events, and the second is for buildings greater than five stories that are likely to be affected by long period motions from distant earthquakes.

The soil conditions in Dushanbe are quite varied, but mostly include rock and loam deposits. Some locations are complicated by buried soil deposits and water lenses. Ground subsidence is occurring in part of the city, particularly on the left bank of the river, causing damage to buildings.

Figure 10d shows the seismic hazard map for the city of Bishkek. This map was created by considering the geological and geomorphologic conditions in the city [36]. The south part of the city has an expected intensity of VIII and is composed of rock deposits overlaid by loamy deposits of less than 2 meters. Groundwater in this region is located at depths greater than 10 meters. The part of the city with an estimated intensity of IX consists of thick layers of sand, loam, and sand with layers of rock. The groundwater depth here ranges from 4 to 10 meters. The northern part of the city has groundwater depth of about 5 meters, with soil deposits of loam, sand, silt, and peat.

The city of Tashkent is composed of three types of soil conditions, which differ in the thickness of Quaternary deposits, the level of groundwater, and the type of underlying deposits. The first type is an average soil condition (type II in the building regulations), which consists of 40 meter thick loess deposits on top of rock with groundwater at a depth of 6 to 20 meters. The second type of soil condition (type III in the building regulations) has 30 to 60 meter thick loess deposits with groundwater at depths of about 25 meters. The third type of soil condition (type I in the building regulations) consists of rock layers of 250 to 300 meters thickness with various degrees of water and sand lenses. The seismic zoning map for Tashkent is shown in Figure 10e and includes zones of expected intensity VIII (soil types I and III) and IX (soil type II) [37].

10. Assessment of Shaking Intensity Potential in Capital Cities

As shown in Figures 10a through 10e, the seismic hazard in the capital cities is characterized by a maximum expected intensity of shaking. It is also desirable to include the potential for shaking due to smaller earthquake events that may also cause extensive damage and loss. The probability of shaking is needed, not only in terms of

I

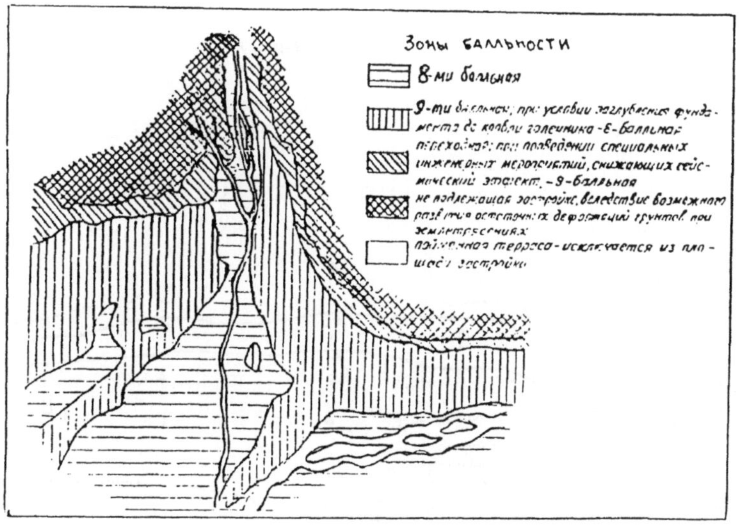

II

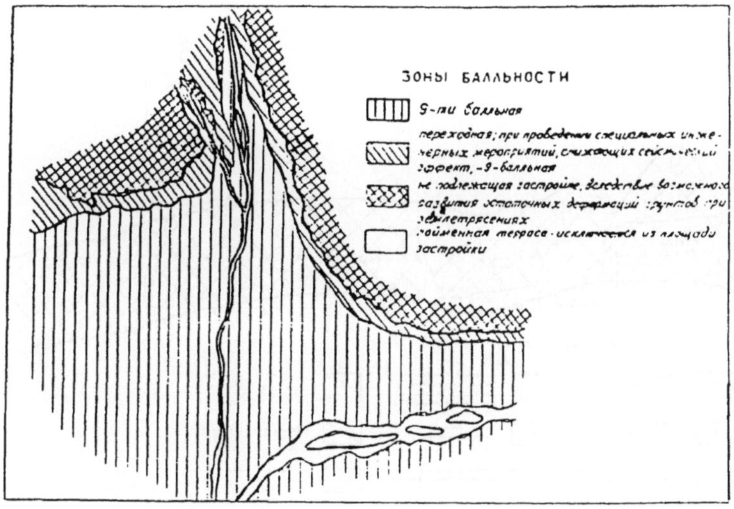

Figure 10c. Map of seismic microzoning for Dushanbe. I - for buildings of five stories or less; II - for buildings greater than five stories tall [35].

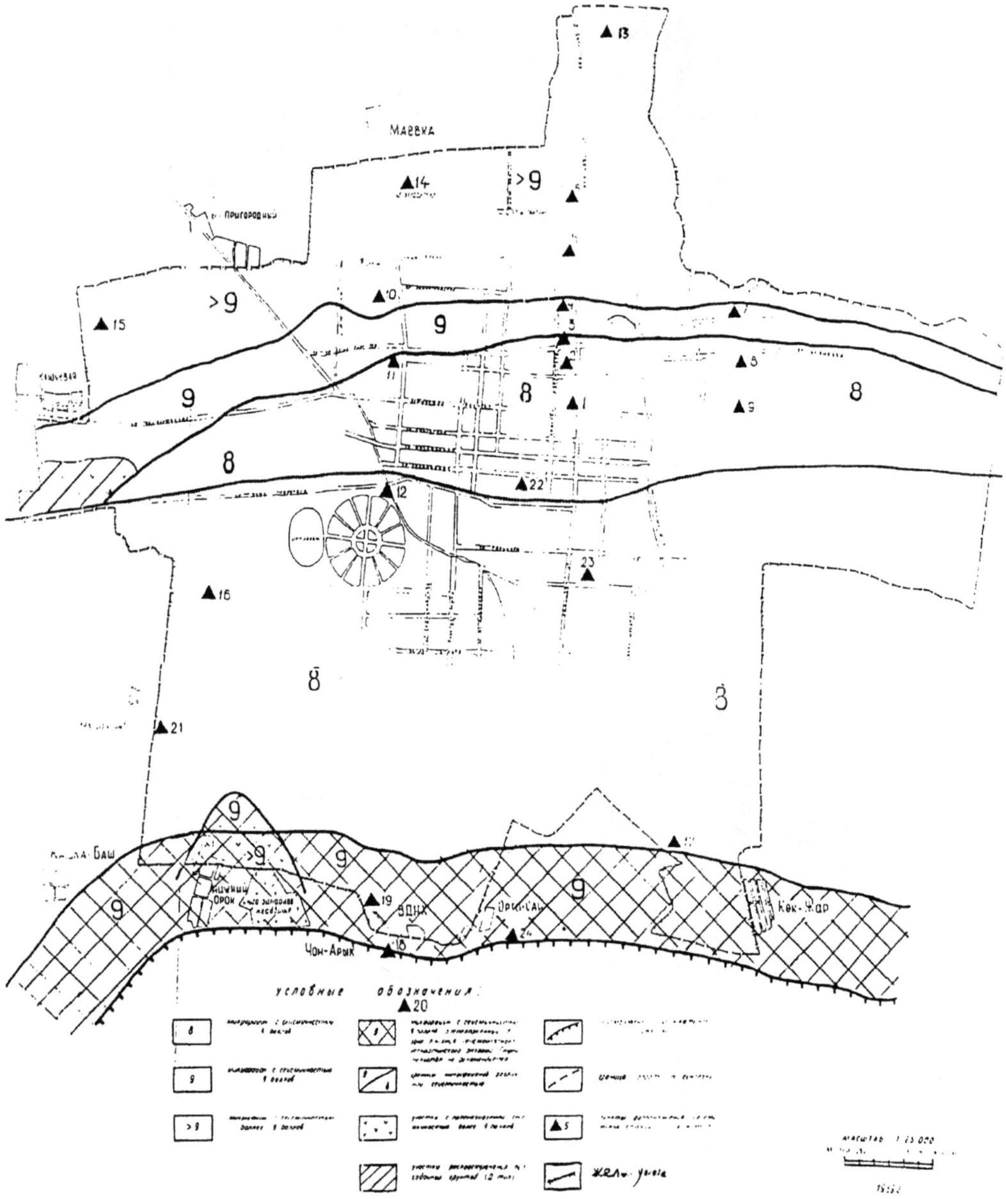

Figure 10d. Map of seismic microzoning of Bishkek [36].

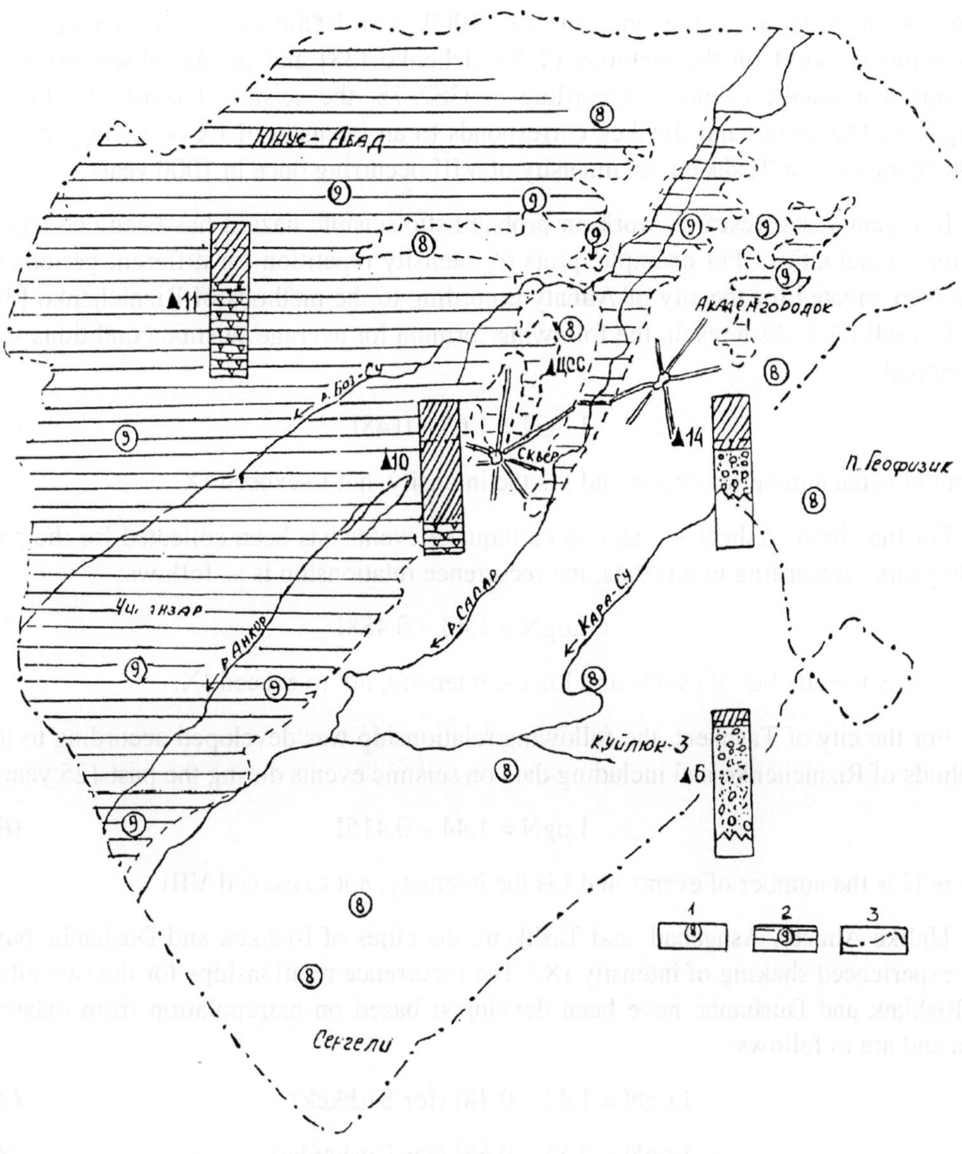

Figure 10e. Map of seismic microzoning of Tashkent [37].

intensity, but also in useful ground motion parameters such as acceleration, velocity, and displacement in both the time and frequency domains.

On the GSZ-78 map, probability is included in only an approximate manner, in terms of an event occurring once in 100, 1000, and 10000 years. This recurrence estimation is based on the methods of Riznichenko [38] and on the observations of previous earthquake events. According to GSZ-78, the seismic hazard of Almaty, Ashgabad, Dushanbe, and Bishkek corresponds to an intensity of IX occurring once in 1000 years, and for Tashkent, an intensity of VIII occurring once in 1000 years.

In recent years, research work on probabilistic seismic hazard has been conducted for the capital cities. For example, plots of intensity repetition for different periods of time were created for the city of Almaty according to the methods of Riznichenko [38] and Cornell [39]. As a result, the following formula for average seismic conditions was developed:

$$\mathrm{Log}N = 1.65 - 0.48I \qquad (4)$$

where N is the number of events and I is the intensity, not to exceed IX.

For the city of Ashgabad, data on earthquake events has been collected for the past 4000 years. According to this data, the recurrence relationship is as follows:

$$\mathrm{Log}N = 1.43 - 0.488I \qquad (5)$$

where N is the number of events and I is the intensity, not to exceed IX.

For the city of Tashkent, the following relationship was developed according to the methods of Riznichenko and including data on seismic events during the past 125 years:

$$\mathrm{Log}N = 1.44 - 0.415I \qquad (6)$$

where N is the number of events and I is the intensity, not to exceed VIII.

Unlike Almaty, Ashgabad, and Tashkent, the cities of Bishkek and Dushanbe have not experienced shaking of intensity IX. The recurrence relationships for the two cities of Bishkek and Dushanbe have been developed based on extrapolation from existing data and are as follows:

$$\mathrm{Log}N = 1.42 - 0.48I \text{ (for Bishkek)} \qquad (7)$$

$$\mathrm{Log}N = 2.55 - 0.65I \text{ (for Dushanbe)} \qquad (8)$$

where N is the number of events and I is the intensity, not to exceed IX.

The return periods for shaking of intensity VII, VIII, and IX for the capital cities were computed according to the relationships given in Equations 4 through 8. The results are shown in Table 6.

Using the return period values in Table 6, it is also possible to compute the exceedance probabilities of various seismic intensity levels for a given future time period. Assuming a Poisson distribution of events, the 50-year exceedance probabilities for seismic intensity in the capital cities are computed as shown in Table 7.

The results shown in Tables 6 and 7 are preliminary; however, they show significant differences in the seismic hazard of cities that have the same level of hazard on the GSZ-78 map. For example, the intensity IX 50-year exceedance probability of Almaty is about 5 times that of Dushanbe and about double that of Ashgabad and Bishkek. On the seismic hazard map for Tashkent, the intensity VIII return period has been reduced from 1000 years to 100 years.

TABLE 6. Return periods of seismic intensity for capital cities of Central Asian republics

City	Return Period (years)		
	Intensity VII	Intensity VIII	Intensity IX
Almaty	51	155	468
Ashgabad	97	298	916
Bishkek	87	263	794
Dushanbe	100	447	1995
Tashkent	29	76	

TABLE 7. 50-year exceedance probabilities of seismic intensity for capital cities of Central Asian republics

City	50-year Exceedance Probability		
	Intensity VII	Intensity VIII	Intensity IX
Almaty	0.62	0.27	0.101
Ashgabad	0.40	0.15	0.053
Bishkek	0.44	0.17	0.061
Dushanbe	0.39	0.10	0.024
Tashkent	0.82	0.48	---

Probabilistic seismic hazard analysis is important for computing earthquake loading criteria in design and retrofit regulations. Currently, there is no consensus as to what the hazard level, i.e., return period or exceedance probability, should be in the building codes. Different seismic design levels have been proposed including the 1000-year return period event, the 100-year return period event, and the event corresponding to 10% chance of exceedance in 50 years. This problem needs to be resolved.

In addition to the question of what should be the design hazard level, the relationship between intensity and physical units of ground shaking needs to be addressed. The current relationship is based on worldwide seismic data and may not be applicable to local regions.

11. Summary

The material presented in this chapter shows that the majority of Central Asia is located in regions with potential for seismic intensity of VII or greater. The capital cities of the five republics are located in zones with potential intensity of VIII and IX. In addition to the estimation of seismic intensity, expected ground motion parameters are needed for the regions of the capital cities. To address this and other issues regarding seismic hazard analysis in the republics of Central Asia, research is needed in the following areas:

- Identification and characterization of seismic source zones that are capable of large earthquake events
- Development of regional attenuation laws for physical ground motion parameters such as acceleration and velocity
- Development of relationships between seismic intensity and physical ground motion parameters such as acceleration and velocity
- Assessment of probabilistic ground motion parameters for defining seismic loading criteria in building codes

12. References

1. Zhunusov, T. et al. (1998) Seismic Hazard and Building Vulnerability in Kazakhstan, in S.A. King, V.I. Khalturin, and B.E. Tucker (eds.), *Seismic Hazard and Building Vulnerability in Post-Soviet Central Asian Republics*, Kluwer Academic Publishers, Dordrecht, pp. 67-92.

2. Imanbekov, S. et al. (1998) Seismic Hazard and Building Vulnerability in Kyrgyzstan, in S.A. King, V.I. Khalturin, and B.E. Tucker (eds.), *Seismic Hazard and Building Vulnerability in Post-Soviet Central Asian Republics*, Kluwer Academic Publishers, Dordrecht, pp. 93-105.

3. Negmatullaev, S. et al. (1998) Seismic Hazard and Building Vulnerability in Tajikistan, in S.A. King, V.I. Khalturin, and B.E. Tucker (eds.), *Seismic Hazard and Building Vulnerability in Post-Soviet Central Asian Republics*, Kluwer Academic Publishers, Dordrecht, pp. 107-125.

4. Karryev, B. et al. (1998) Seismic Hazard and Building Vulnerability in Turkmenistan, S.A. King, V.I. Khalturin, and B.E. Tucker (eds.), *Seismic Hazard and Building Vulnerability in Post-Soviet Central Asian Republics*, Kluwer Academic Publishers, Dordrecht, pp. 127-145.

5. Rashidov, T. et al. (1998) Seismic Hazard and Building Vulnerability in Uzbekistan, in S.A. King, V.I. Khalturin, and B.E. Tucker (eds.), *Seismic Hazard and Building Vulnerability in Post-Soviet Central Asian Republics*, Kluwer Academic Publishers, Dordrecht, pp. 147-180.

6. Rantsman, E.Y. (1979) *Places of Earthquakes and Morphological Structure of Mountain Countries*, Nauka, Moscow.

7. Tal-Virsky, B.B. (1982) *Geographical Fields and Tectonics of Central Asia*, Nedra, Moscow.

8. Toksöz and Nafi (1975) The Subduction of the Lithosphere, *Scientific American* **11**.

9. Krestnikov, V.N. et al. (1979) *Quaternary Tectonics of Pamirs and Tien-Shan*, Nauka, Moscow.

10 Krestnikov, V.N. and Reicner, G.I. (1967) Newest Tectonics of East Kazakhstan and Djungaria, *Geotectonics* **2**, 71-87.

11 Rustanovich, D.N. (1967) *Seismicity of the Turkmen SSR Territory and Ashgabad Earthquake of 1948*, Nauka, Moscow.

12 Kondorskaya, N.V. and Shebalin, N.V. (1977) *New Catalogue of Large Earthquakes on the Territory of USSR*, Nauka, Moscow.

13 Vedenskaya, A.V. (1969) *Research of Strains and Breaches in Earthquakes Focuses by Means of the Theory of Dislocations*, Nauka, Moscow.

14. Lukk, A.A. et al. (1981) Seismic Tectonic Deformation of Central Asia and Kazakhstan, in *Earthquakes of Central Asia and Kazakhstan*, Donish, Dushanbe, 67-100.

15. Gubin, I.E. (1960) *Regularity of Seismic Events in the Territory of Tajikistan*, USSR Academy of Science, Moscow.

16. _____ (1986) *Gazli Earthquakes of 1976 and 1984*, Fan, Tashkent.

17. Nurmagambetov et al. (1979) Earthquake on the 25th of September 1979, in *Earthquakes in the USSR in 1979*, Nauka, Moscow, 48-53.

18. Lukk, A.A. (1983) Seismic-Tectonic Deformation and Strained State of the Earth Crust of the Central Asia and Kazakhstan, in *Earthquakes of Central Asia and Kazakhstan*, Donish, Dushanbe, 118-135.

19. Yunga, S.L. et al. (1993) Mechanisms of Earthquakes Focuses and Seismic-Tectonic Deformation of Central Asia Earth Crust, in *Earthquakes in the USSR in 1989*, Nauka, Moscow, 225-235.

20. Nazarov, A.G. and N. V. Shebalin, N.V. (1975) *Seismic Scale and Methods of Seismic Intensity Measuring*, Nauka, Moscow.

21. _____ (1979) *Seismic Zoning of Kazakhstan*, Nauka, Almaty.

22. _____ (1988) *Strong Motions in Earthquakes*, Donish, Dushanbe.

23. Aptikayev, F.F. and Kopnichev, Y.F. (1979) . Taking into Account Earthquake Focus Mechanism in the Prediction of Strong Motion Parameters, *DAN T 247* **4**, 822-825.

24. Fukyshima, Y. and Tanaka, T. (1990) A New Attenuation Relation for Peak Horizontal Acceleration of Strong Earthquake Ground Motion in Japan, *BSSA* **4**, 757-783.

25. _____ (1979) *Methodical Recommendations for Seismic Zoning of the USSR Territory*, IFZ AN USSR, Moscow.

26. _____ (1990) *Seismic Zoning of the USSR Territory*, Nauka, Moscow.

27. Aptikayev, F.F. and Knauf, V.I. (1986) Methodical Recommendations on GSZ, *Issues of Engineering Seismology* **27**, 184-211.

28. _____ (1984) USSR Seismic Zoning Map Scale 1:5,000,000, Nauka, Moscow.

29. Medyedev, S.V. International Scale of Seismic Intensity, in *Seismic Zoning of the USSR*, Nauka, Moscow, 151-162.

30. _____ (1971) Tashkent Earthquake on the 26th of April 1966, FAN, Tashkent.

31. _____ (1994) Detailed Seismic Zoning Experience of Almaty Industrial Zone, in *Problem of Earthquakes and Seismic Hazard Prediction*, 130-136.

32. _____ (1988) *Detailed Seismic Zoning of East Kyrgyzstan*, Ilim, Frunze.

33. _____ (1994) *Construction of Almaty and Surrounding Territories Taking into Account Seismic Micro Zoning*, Almaty.

34. Vatolin, B.S. and Esenov, E.M. (1984) *Seismic Microzoning Map of Ashgabad Territory.*

35. Negmatullayev, S.H. et al. (1987) *Tajikistan Service of Strong Motions*, Donish, Dushanbe.

36. Turdukulov, A.T. et al. (1989) *Schematic Map of Complex Seismic Microzoning of Frunze Territory.*

37. Kasimov, S.M. (1984) *Seismic Microzoning Map of Tashkent.*

38. Riznichenko, Y.V. (1979) *Seismic Shaking of the USSR Territory*, Nauka, Moscow.

39. Cornell, C.A. (1968) Engineering Seismic Risk Analysis, *BSSA* **5**, 1583-1606.

40. Aptekman, Z.Y. et al. (1978) Some Data on Processes in Epicentral Zone of Gazli Earthquakes of 1976, in *Epicenter Zone of Earthquakes*, Nauka, Moscow, 149-170.

EARTHQUAKE RESISTANCE OF MULTI-STORY RESIDENTIAL BUILDINGS IN CENTRAL ASIAN CAPITAL CITIES

S. IMANBEKOV
Seismic-Resistant Construction Institute, Kyrgyzstan
S. URANOVA
Seismic-Resistant Construction Institute, Kyrgyzstan
W. IWAN
California Institute of Technology, USA

1. Introduction

Development in the capital cities of the Central Asian republics has been carried out under the unified structure of the construction industry with a strict centralized system for developing and approving building type design and construction codes. Because of this, residential buildings of the same type in these cities are in general of identical design and construction.

The seismic vulnerability of a building or other structure may be defined as its susceptibility to damage during an earthquake having a specified level of ground shaking. This damage may result in physical injury or death to occupants, temporary or permanent loss of function of the building, and associated economic impact. The degree of seismic vulnerability of a building will depend upon its general configuration, load transfer system, design specifics, and the quality of materials and construction. By carefully examining these factors, it is possible to estimate the level of vulnerability of a class of buildings or a specific building structure.

Nine prevalent types of residential buildings in urban areas of Central Asia are shown in Table 1. The most common residential buildings in the five capital cities of the Central Asian republics can be divided into the five following types: buildings with brick walls, large-panel buildings, buildings with monolithic reinforced concrete walls, buildings with reinforced concrete frame, and buildings with flexible first floors. Each of these is described in more detail in the next section. The information in this chapter is derived from reports written for the NATO Advanced Research Workshop on "Strategies for Seismic Risk Reduction in Urban Territories of Central Asia" held in October of 1996. These reports are included in this book.

S.A. King et al. (eds.),
Seismic Hazard and Building Vulnerability in Post-Soviet Central Asian Republics, 45–65.

TABLE 1. Residential building types in urban areas of Central Asian republics

Building Type	Description
1	Adobe buildings
2	Brick bearing wall systems with wooden floors, 1-2 stories, pre-1955
3	Brick bearing wall systems with pre-cast reinforced concrete (R/C) floor panels, 3-5 stories, pre-1957
4	Brick bearing wall systems with pre-cast reinforced concrete (R/C) floor panels, post-1957 (earthquake code incorporated dynamic action)
5	Pre-cast R/C frame systems, 4-9 stories with brick infill walls, and welded joints (highly variable quality of construction)
6	Pre-cast R/C frame systems having *weak first stories*, 4-9 stories with brick infill walls and welded joints (highly variable quality of construction)
7	Mixed systems with pre-cast R/C frame systems, 4-9 stories with pre-cast R/C panels (highly variable quality of construction)
8	Pre-cast R/C large panel systems with dry or wet joints (including Series 464)
9	Other systems including block systems, flat slab (no beams) systems, and lift-slab systems, buildings on poor (collapsible soils)

2. Typical Residential Building Types in Central Asian Cities

Table 2 lists the distribution of residential building types in the Central Asian republics by capital city. The design and construction of these building types, as well as their vulnerability to seismic loads, are described in this section.

2.1 BUILDINGS WITH BRICK WALLS

Buildings with brick walls (see Figures 1 and 2) are one of the most vulnerable building types in terms of seismic resistance. The design and construction of the primary load bearing elements depend on the code that was in enforcement at the time of construction. A considerable number of buildings are quite old and were constructed before seismic load criteria were added to design codes.

All design code requirements for buildings with brick walls can be divided into the following four groups:

1. Requirements for materials specify that brick must be of grade 75 or higher and the tensile resistance of brick wall joints must be 120 kPa or higher

Figure 1. Typical brick wall building construction in Central Asia.

Figure 2. Typical brick wall building construction in Central Asia.

2. Restrictions are given for dimensions of the building plan, distance between walls, height of floors, and sizes of columns and windows
3. Requirements are given for the construction of reinforced concrete floors, including the reinforcement at the floor level and the connections between the floors and the walls
4. Use of reinforced concrete supports is encouraged, as long as the dimension requirements are met

TABLE 2. Statistics of the most common types of residential buildings in capital cities of Central Asia

Capital City	Type of Buildings	Number of Buildings	Square Footage (x 1000 sq. m.)	Occupants (thousands)
Almaty, Kazakhstan	Brick walls	4150	6580	530
	Flexible first floor	230	350	35
	Concrete frame	230	350	35
	Large-panel	2755	5550	400
Bishkek, Kyrgyzstan	Brick walls	1060	1446	100
	Flexible first floor	20	30	5
	Concrete frame	67	105	8
	Large-panel	1020	3200	213
Dushanbe, Tajikistan	Brick walls	1000	1440	120
	Flexible first floor	-	-	-
	Concrete frame	25	30	2
	Large-panel	1500	4570	360
Ashgabad, Turkmenistan	Brick walls	917	1100	100
	Flexible first floor	-	-	-
	Concrete frame	20	25	2
	Large-panel	840	2540	200
Tashkent, Uzbekistan	Brick walls	4200	6750	541
	Flexible first floor	---	---	---
	Concrete frame	2700	3250	247
	Large-panel	4000	11725	699

Due to the low quality of construction work, most of the brick buildings do not provide adequate earthquake resistance. In general, the brick and mortar are inadequate, as are the additional reinforced concrete supports. Design specifications for reinforcement of brick walls and reinforced concrete elements are not being followed. Results from recent large earthquakes have shown that the rate of damage to brick wall buildings can be as high as 80%, which is drastically higher than the rate assumed in the design codes.

2.2 LARGE-PANEL BUILDINGS

The primary load carrying elements in a large-panel buildings are large reinforced concrete panels; there is no frame (see Figure 3). The panels are connected by concrete joints or steel details, as shown in Figures 4 and 5. The interior of the panels is made of hard concrete, while the exterior is made of light concrete or light concrete over a layer of hard concrete. The concrete layers are typically 12-16 cm thick. These buildings are considered to have the highest seismic resistance, and only experience problems with corrosion of the joint elements.

2.3 BUILDINGS WITH MONOLITHIC REINFORCED CONCRETE WALLS

Buildings of this type have no frame (see Figure 6). The thickness of the interior walls is typically 12-16 cm. The thickness of the exterior walls is determined according to heating requirements. These buildings are assumed to have adequate seismic resistance. Buildings with moveable casing have been observed to experience more earthquake damage than those with sliding casing.

2.4 BUILDINGS WITH REINFORCED CONCRETE FRAME

The reinforced concrete frames are either monolithic or precast of linear elements (see Figures 7 and 8). The earthquake resistance of this type of building depends heavily on the design and construction. The buildings are divided into the following three groups depending on the way in which the frame carries the seismic load:

- Frame with rigid connections of columns and beams
- Frame with some diagonal bracing
- Braced frame

A substantial number of buildings with precast reinforced concrete frames are constructed of linear elements with reinforcement connected by welds. The low seismic resistance of these buildings is due to the location of joints in the areas of greatest load, and the low load-bearing capacity of elements.

Frame buildings without beams are very vulnerable to seismic loads. The total destruction of these buildings in the Spitak, Armenia earthquake is attributed to the poor design and construction of connections between floor slabs and columns, and the extreme differences in rigidity between the floors and columns.

Figure 3. Typical large-panel building construction in Central Asia.

Figure 4. Typical large-panel building construction in Central Asia.

Figure 5. Typical large-panel building construction in Central Asia.

Figure 6. Typical reinforced concrete wall building construction in Central Asia.

Figure 7. Typical reinforced concrete frame building construction in Central Asia.

Figure 8. Typical reinforced concrete frame building construction in Central Asia.

2.5 BUILDINGS WITH FLEXIBLE FIRST FLOORS

In this type of building, the ground floor is typically a reinforced concrete frame without any stiffening diaphragm. The upper floors are supported by brick walls or large panels. The high vulnerability to earthquake damage of this type of building is well known throughout the world.

2.6 ADOBE COTTAGES

Most one- and two-story cottages fall under the category of regional peculiarities of residential buildings in the capital cities of the Central Asian republics. In some cities, 30 to 40% of the population live in such cottages. In general, these cottages are constructed of available materials without any consideration of seismic loads (see Figures 9 and 10).

3. A Damage Scale for Residential Buildings in Central Asia

The vulnerability of different types of residential buildings in urban areas of the Central Asian republics, may be specified in terms of a damage scale that consists of five degrees of damage. The five damage degrees are listed in Table 3. Damage degrees 1-2 are repairable while damage degrees 4-5 are not economically repairable. Damage degree 3 represents the intermediate damage state between clearly repairable and clearly not repairable damage. The threat to life safety generally begins at damage degree 4.

TABLE 3. Damage scale for residential buildings in Central Asia

Damage Level	Damage Description	Reparability
Degree 1	Slight	Repairable
Degree 2	Moderate	↑
Degree 3	Heavy	----------------------------
Degree 4	Partial collapse	↓
Degree 5	Total collapse	Not economic to repair

Figure 9. Typical adobe cottage residential construction in Central Asia.

Figure 10. Typical adobe cottage residential construction in Central Asia.

4. Vulnerability of Residential Buildings to Seismic Loads

The low level of seismic resistance of the existing residential buildings can be attributed to the following factors:

- Many buildings are old and were built according to design and construction codes that do not have seismic load considerations; in addition, the buildings have suffered deterioration due to wear and tear and decreasing strength of structural elements and connections
- In the current design codes, the reliability level is not the same for all types of buildings
- Design and construction specifications are not followed, resulting in poor construction quality
- Underestimation of the seismic hazard at the building site results in a design that is not adequate for seismic loads
- Changes in the use of a building or an addition to the building can cause the load bearing capacity of the structural elements to be exceeded
- Ground settlement over time results in cracked foundations and misalignment of structural elements
- Inadequate design and construction codes are still used due to the long time it takes for the industry to change them
- Certain building types were constructed in large quantities without adequate testing of resistance to large earthquake forces
- Many specialists have left design companies and current staff members have not been trained in proper design techniques
- There is a lack of new information about earthquake resistant construction and engineering seismology

Table 4 indicates the consensus vulnerability of different types of residential buildings in Central Asia. Vulnerability is expressed in terms of damage level and is given as a function of building type and MSK intensity.

It is evident from Table 4 that an intensity IX event would cause collapse or very serious (not repairable) damage to most buildings in category types 1-6. In the Central Asian Republics it is generally believed that reinforced concrete large panel buildings (type 8) will perform better than other types of residential buildings. However, even the damage to these structures will be moderate to heavy.

TABLE 4. Consensus vulnerability of residential buildings in Central Asia (damage levels corresponding to various buildings types and intensity)

	MSK Intensity			
Building Type (see Table 1)	VI	VII	VIII	IX
1 - Adobe	1-2	3	4-5	5
2 - Brick/Wood	1-2	2-3	4	5
3 - Brick – precast, pre-1957	0-1	1-2	3-4	4
4 - Brick – precast, post-1957	0	0-1	2-3	3-4
5 - R/C – precast frame	0	1	2-3	3-4
6 - R/C – flexible first floor	0	1	3	3-5
8 - Large-panel (S-464)	0	0-1 (1)	1-2 (2)	2 (3)

5. Building Codes and Regulations for Seismic Resistance

A unified system of building regulations by the construction industry (SNiP) has been functioning in the republics of Central Asia since before 1992. Regulations specific to each republic were reflected in the republic norms (RSN). The regulations have changed over time to reflect the improvements in earthquake resistant design, and changes in building dimension requirements. Table 5 lists the building codes that have been in place in the Central Asian republics.

In the first code (PSP), design of buildings was based on static theory. Seismic design according to dynamic theory was introduced in the codes SNiP II-A.12-62 and SNiP II-A.12-69. In these specifications, seismic loads for structures were computed based on ratios reflecting the intensity and dynamic effects of an earthquake.

Methods for determining seismic loads were further developed in SNiP II-7-81*. For the time (1981), this regulation was quite progressive in that it contained a number of new provisions for determining seismic loads for buildings and other structures. The general dynamic theory was not changed, but additions were made to reflect new developments in the theory and practice of earthquake resistant design. The most important changes include the following:

- Considering the impact of ground conditions on seismic loads – by including dynamic ratios for characterizing the effects of soft, moderate, and hard soils
- Determining the seismic load according to actual levels of ground accelerations during earthquake events – assuming elastic behavior of the structure

- Considering the recurrence intervals for earthquake events – including amplitude of shaking, duration, and dominant periods of motion; design specifications are made based on reliability and seismic risk theory

TABLE 5. List of building codes in the republics of Central Asia

Building Code	Time Period	Territory
PSP 101-51. Regulations of building in seismic regions	1951-1957	All Central Asian republics
SN-8-57. Building norms and guiding principles in seismic regions	1957-1962	All Central Asian republics
SNiP II-A.12-62. Building in seismic regions: Design codes	1963-1970	All Central Asian republics
SNiP II-A.12-69. Building in seismic regions: Design codes	1970-1977	All Central Asian republics
SNiP II-A.12-69*. Building in seismic regions: Design codes	1977-1982	All Central Asian republics
SNiP II-A.12-81. Building in seismic regions: Design codes	1982-1991	All Central Asian republics
SNiP II-A.12-81*. Building in seismic regions: Design codes	1991-1996	All Central Asian republics
KMK 2.01.03-96. Building in seismic regions: Design codes	1996	Uzbekistan
RSN 10-70. Building of Alma-Ata city and adjoining territories taking into account seismic microzoning	1983-1995	Kazakhstan
SN RK B.2.2-7-95. Building of Alma-Ata city and adjoining territories taking into account seismic microzoning	1995	Kazakhstan
SNiP 2.01.01-93 KR. Building on the territory of Bishkek taking into account seismic microzoning and ground conditions	1993	Kyrgyzstan
SNiP 2.01.02-94 KR. Building in Kyrgyzstan with seismicity in the region of more than 9	1994	Kyrgyzstan
SNiP 31-01-95 KR. Re-function same apartment of existing residential buildings	1995	Kyrgyzstan

6. Reducing Potential Damage from Earthquakes in Central Asian Republics

The following techniques have been used for the rehabilitation of buildings depending on the technical status of the project:

- Without changing the existing design scheme of a building – rehabilitation includes the reinforcement of existing elements and the links among them
- With partial alteration of the existing design scheme of a building – rehabilitation includes improvements to the spatial layout such as adding longitudinal and transverse walls and seismic joints
- Changing of the design scheme and dynamic characteristics of a building – rehabilitation includes adding rigid diaphragms, fixing links, and other measures as described below for various building types:
    - For brick and masonry buildings – gunite of walls and multi-layer covering metal grids with high grade grout; construction of metal and reinforced concrete jackets for narrow columns; addition of prestressed steel reinforcement; construction of specific ties and links among structural elements
    - For large-panel buildings – reinforcement of connections by adding pre-stressed steel belts and dowels and polymer-coated steel
    - For frame buildings – epoxy injection of cracks; addition of vertical and horizontal steel braces; construction of reinforced concrete or steel jackets for narrow columns and beams; addition of reinforcement layers to floors

The major barriers to rehabilitation of existing residential buildings in the republics of Central Asia are as follows:

- Lack of financing
- Lack of regulations or codes for rehabilitation of buildings
- Lack of understanding of how added reinforcing elements interact with existing structural elements
- Limited quantity of necessary construction materials
- Limited experience of design professionals in seismic rehabilitation practices
- Lack of legislation that would mandate increasing the seismic resistance of certain types of buildings
- Lack of necessary technological tools for observing building performance

The following actions would help to solve some of the barriers to seismic rehabilitation listed above:

- Identify sources of funding

- Implement procedures for inspection and certification of apartment buildings to identify the most vulnerable buildings
- Analyze available information on seismic rehabilitation practices and develop new ones if needed, including modern technology such as base isolation and composite materials
- Develop schedules with local governments for rehabilitating vulnerable buildings
- Develop design regulations for rehabilitation of various types of buildings
- Complete rehabilitation of one building of each type and test to determine how well the retrofit works

Costs for seismic rehabilitation of the most vulnerable types of buildings are estimated to be about 25-40% of the cost of new construction, or currently about 80-160 US$ per square meter. This is assuming about 25-40 kg of steel, 40-60 kg of concrete, and 1-3 man-days per square meter. According to the building statistics for each capital city, the total rehabilitation costs for vulnerable buildings is estimated at 340 million US$ for Tashkent, 300 million US$ for Almaty, 208 million US$ for Bishkek, 150 million US$ for Dushanbe, and 160 million US$ for Ashgabad.

As a first step, the most vulnerable building type of 2-4 stories in height should be identified in each republic, and rehabilitation techniques should be developed. For example, in Uzbekistan it could be buildings of the 210 series, in Kazakhstan and Kyrgyzstan, buildings of series 308, in Tajikistan, buildings of series TJ-401, and in Turkmenistan, 1-295 series buildings. Such a procedure could be implemented over a time period of one to two years, at a cost of about 2 to 3 million US$ per republic.

7. International Cooperation for Preventing Damage from Earthquakes

Considering the high seismic vulnerability of buildings in the republics of Central Asia, international cooperation would help to reduce potential damage from future earthquakes. Cooperation should be arranged in the development of procedures for seismic rehabilitation of buildings, staff training, and exchange of scientific and technical information on seismic hazard and risk analysis techniques. Potential forms of international cooperation include the following:

- Development of joint research projects on preventing potential earthquake damage
- Creation of a unified system for training scientific staff on issues of earthquake resistant construction, reliability of buildings for seismic loads, engineering seismology, seismic risk analysis, writing of design codes for earthquake resistant construction, and organization and financing of rehabilitation projects
- Creation of a central information center on earthquake issues, including collecting and distributing all information (periodicals, design and construction codes,

software, and monographs) on earthquake-resistant construction, seismic risk and reliability of buildings, determining seismic loading, and engineering seismology

- Organization of joint meetings and conferences

SEISMIC HAZARD AND BUILDING VULNERABILITY IN KAZAKHSTAN

T. ZHUNUSOV
KazNIISSA, Kazakhstan
A. TAUBAEV
KazNIISSA, Kazakhstan
I. ITSKOV
KazNIISSA, Kazakhstan
N. MIKHAILOVA
KazNIISSA, Kazakhstan
A. NURMAGAMBETOV
UIPE, Complex Seismological Expedition, Kazakhstan

1. Main Data on the Capital City of Kazakhstan

The capital city of the republic of Kazakhstan is Almaty. The old name is Verny. The city is situated at the foot of the Zailiski Alatau range. Geographic coordinates of the central part of the city are 40°14.5' north latitude and 76°56.9' east longitude. The elevation of the city is 700 to 900 meters. The founding date of Kazakhstan is considered to be 1868, the year when the fortress Verny, which was located here, became the city of Verny.

1.1 SHORT GEOMORPHOLOGICAL DESCRIPTION OF ALMATY

The city of Almaty is situated near the north slope of the Zailiski Alatau range on the area of deposits of the Bolshaya and Malaya Almatinka rivers. Within the territory of the city, two main relief complexes exist: (a) mountain, represented by low mountains and a chain of foothills situated at an elevation of 1 to 5 km, and (b) plain, which has a gentle slope of 1° to the North and contains many ravines, gorges and gullies.

1.2 SEISMOTECTONIC DESCRIPTION OF ALMATY

The region of Almaty is situated in the zone joining the Zailiski Alatau range, the region of steady, recent upthrust, with the Ili hollow, the region of steady, recent subsidence. The boundary line dividing these two regions, on which the Almaty territory is located, is under the influence of differentiated slow earth crustal motions.

The Zailiski fault, which is traced along the North slope of the Zailiski Alatau range, presents the greatest seismic hazard for the city of Almaty. Parallel to it and

S.A. King et al. (eds.),
Seismic Hazard and Building Vulnerability in Post-Soviet Central Asian Republics, 67–92.

across the city territory runs the Almatinski fault, and there are also a number of smaller faults. The seismicity of the faults crossing the city territory has not yet been defined. Motions are possible along them due to strong earthquakes with epicenters outside the city. South of the Zailiski fault and parallel to it runs a series of abyssal faults, the Chiliko-kemin series. Figure 1 shows the main active faults defining the seismic hazard of the Almaty region.

1.3 GENERAL SEISMIC ZONING FOR KAZAKHSTAN

The current map of general seismic zoning for Kazakhstan was made in 1978 as part of the total map of seismic zoning for the whole territory of the USSR. The map developed for the USSR is shown in Figure 2. According to this map, the estimated intensity for the city of Almaty is IX.

1.4 TOTAL AREA AND POPULATION OF ALMATY

The total area of Almaty is 300 square km. and the population is roughly 1.5 million people.

1.5 MULTISTORY BUILDINGS IN ALMATY

The total area of multistory buildings in Almaty is about 50 square km., inhabited by about 1 million people.

1.6 DESCRIPTION OF SOIL CONDITIONS IN ALMATY

The central part of Almaty is formed by a thick layer of rocky soil with sand (sometimes sandy loam or loamy soil) filling, covered with thin layers of covering formation (1 to 5 meters thick). Groundwater is located at a depth of 15 to 20 meters. According to the seismic zoning map shown in Figure 2, this region is expected to have intensity IX shaking.

The northern part of the city contains layers of sandy loam, loam, and sands of different coarseness. Below the depth of 20 meters, there are sandy soils with layers of crushed rock. Groundwater is at a depth of 3 to more than 10 meters. Maximum intensity in this region is expected to be IX to X.

In the eastern part of the city, soils are characterized by clay deposits and sandy loam deposits up to 10 meters thick. Below this are pebbles with layers of loam, sandy loam, and sand. Groundwater is located at a depth of 5 to 10 meters. Maximum intensity in this region is estimated to be X.

In the north-west part of the city, there are loam and sandy loam deposits on top of sands. Groundwater in this region is at a depth of 5 to 10 meters. Maximum intensity of VIII is expected in this area. Figure 3 shows a map of detailed microzoning for the city of Almaty.

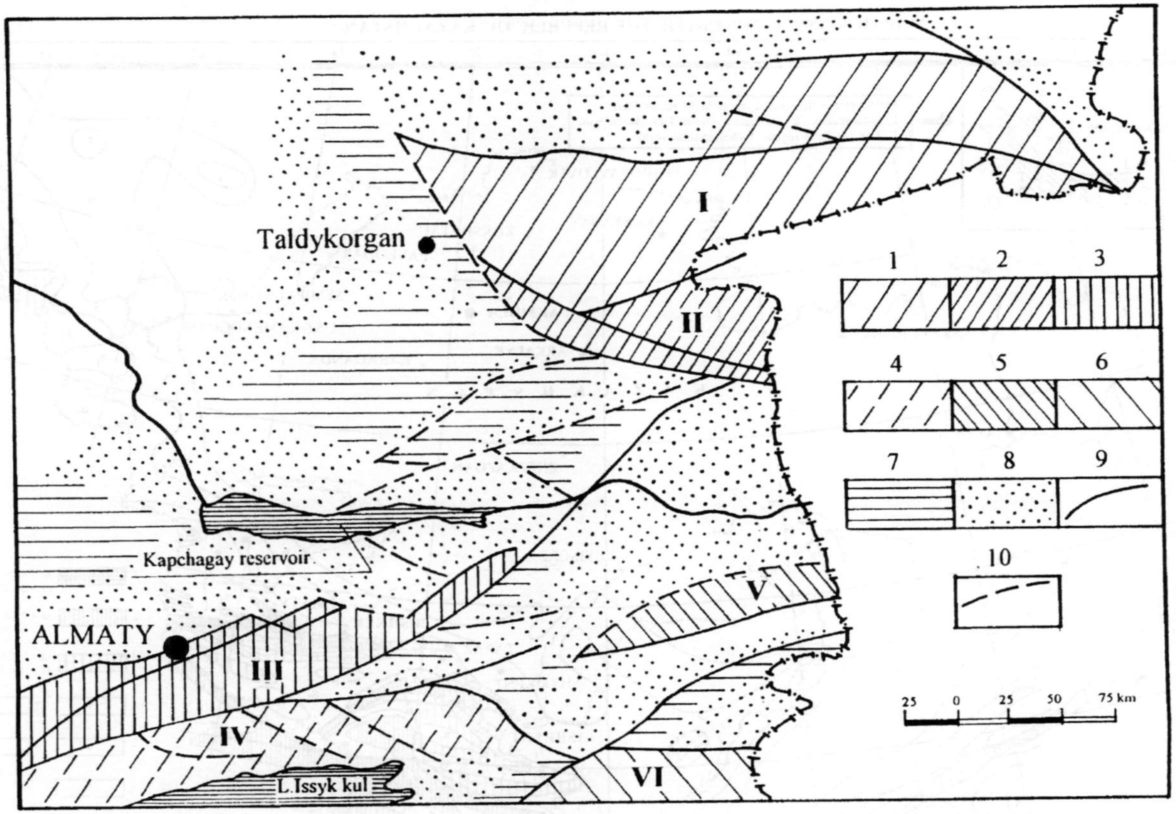

Figure 1. Scheme map of the newest structures of North Tien-Shan and Zhungaria (Patalakha, Chabdarov, 1978). 1-6 = large blocks expressed by high range (1 = Zhungaria, 2 = Zharkent, 3 = Zailiysk, 4 = Kungey, 5 = Terskey); 7 = block expressed by medium and small ranges; 8 = blocks expressed by depression; 9 = regional faults; and 10 = faults of smaller level.

MAP OF SEISMIC ZONING OF THE REPUBLIC OF KAZAKHSTAN

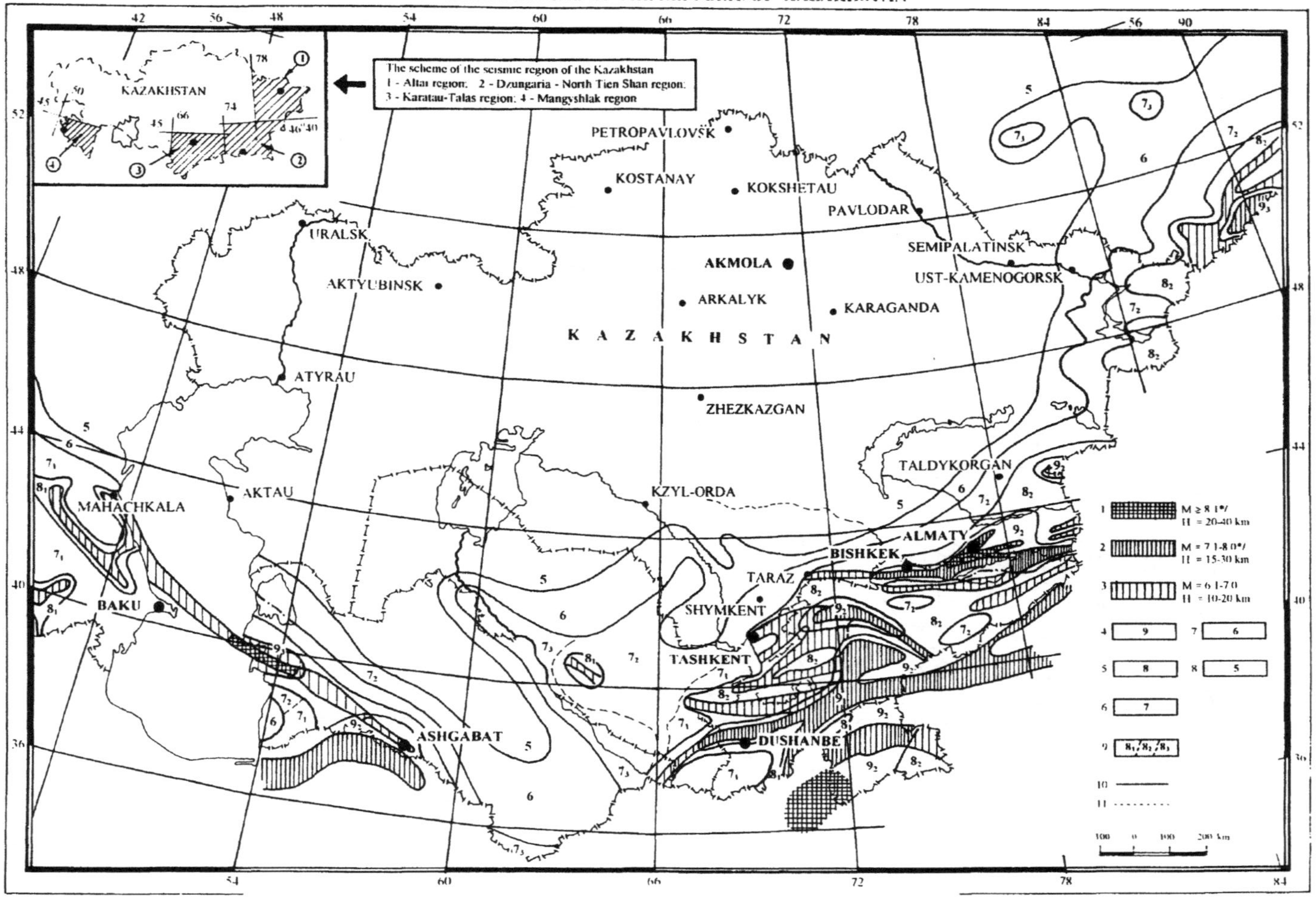

Figure 2. Map of seismic zoning of the Republic of Kazakhstan.

Figure 3. Scheme of complex seismic zoning of Almaty. 1 = Intensity 10; 2 = Intensity 9; 3 = Intensity 8; 4 = areas with possible surface faulting.

2. Characteristics of Seismic Hazard and Expected Intensity in Kazakhstan

Figure 4 shows a portion of the total seismic zoning map (GSZ-78) for the territory surrounding Almaty. Figure 5 shows a portion of the detailed seismic zoning map for the region of Almaty. The seismic generating zones in the area around Almaty are shown in Figure 6 with parameters as shown in Table 1.

TABLE 1. Main parameters of seismic generating zones in Almaty

Zone Number	Zone Name	M_{max}	Depth (km)	Typical Mechanism	LogN = a –bM	
					a	b
1	Kungey	8.2	25	upthrust	2.28	0.61
2	Zailiski	8.0	20	upthrust	1.81	0.63
3	Almaty	7.0	15	upthrust	1.46	0.76
4	South-Issyk kul	7.5	20	upthrust	2.46	0.73

According to the GSZ-78 map, the maximum intensity in Almaty is IX with a recurrence interval of 1000 years. Figure 7 shows the recurrence relationship for Almaty which is described by the following formula:

$$\text{Log N} = 1.65 - 0.48\ \text{I} \qquad \text{for I} \leq \text{IX} \qquad (1)$$

where N is the number of events and I is the intensity on the MSK scale. According to this relationship, intensity VII is expected once in 50 years, intensity VIII is expected once in 150 years, and intensity IX is expected once in 500 years.

The intensity attenuation relationship for the Almaty region is shown in Figure 8 and is represented as follows:

$$I_o = a\ M_s - b\ \text{Log H} + c \qquad (2)$$

where I_o is intensity, M_s is magnitude, H is distance in km., and a, b, and c are parameters. Average parameter values are a = 1.5, b = 3.6, and c = 3.3. Across structures, parameter values are a = 1.5, b = 3.7, and c = 3.1. Along structures, parameter values are a = 1.5, b = 3.3, and c = 2.9.

For attenuation of acceleration in the Almaty region, a world-wide relationship is used, taking into consideration the results of special investigations on the possibility of using the relationship in the region of North Tien-Shan. The relationship was derived by F.F. Aptikayev for earthquakes with M > 5 and is as follows:

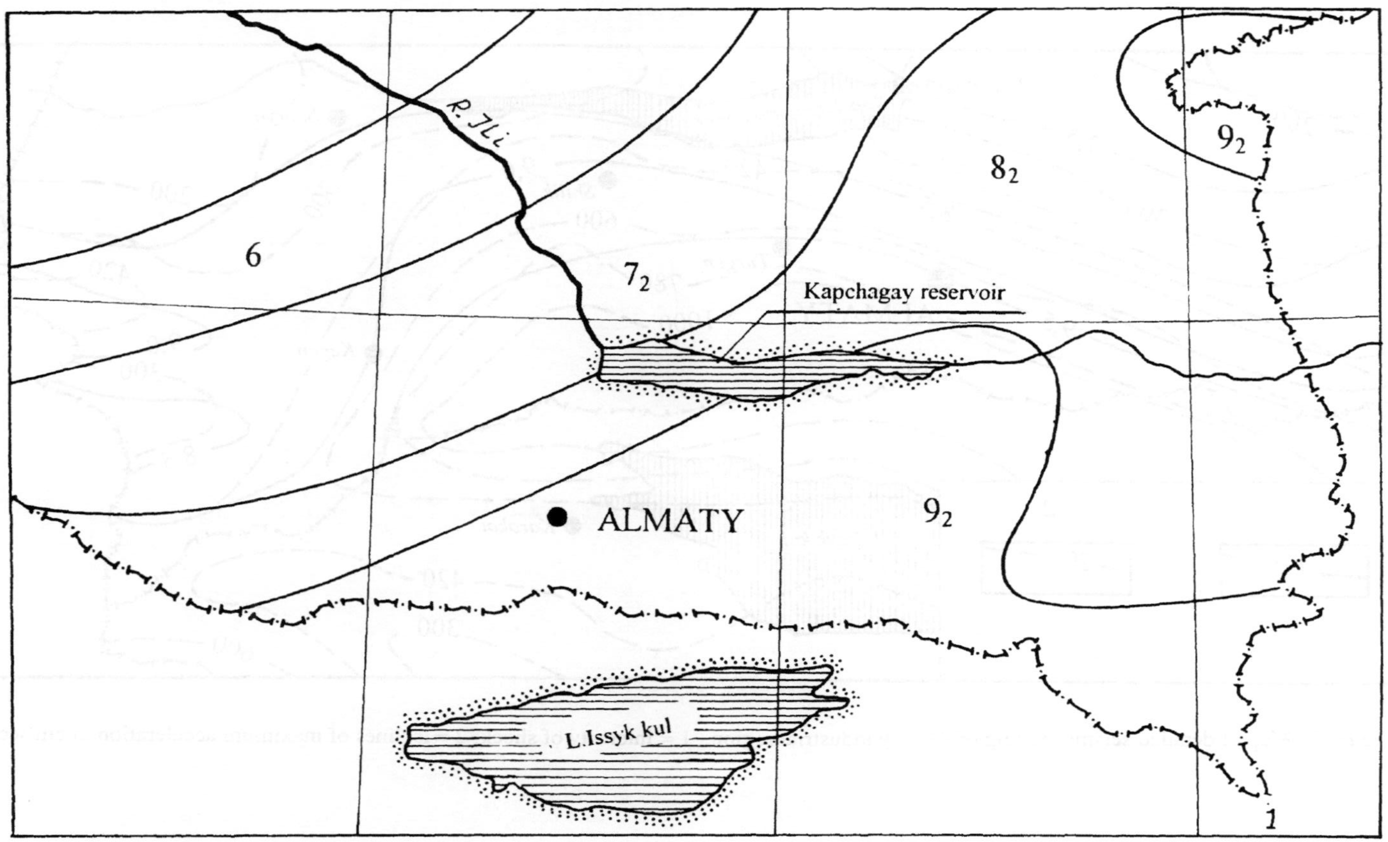

Figure 4. Fragment of the map of general seismic zoning for the Almaty region.

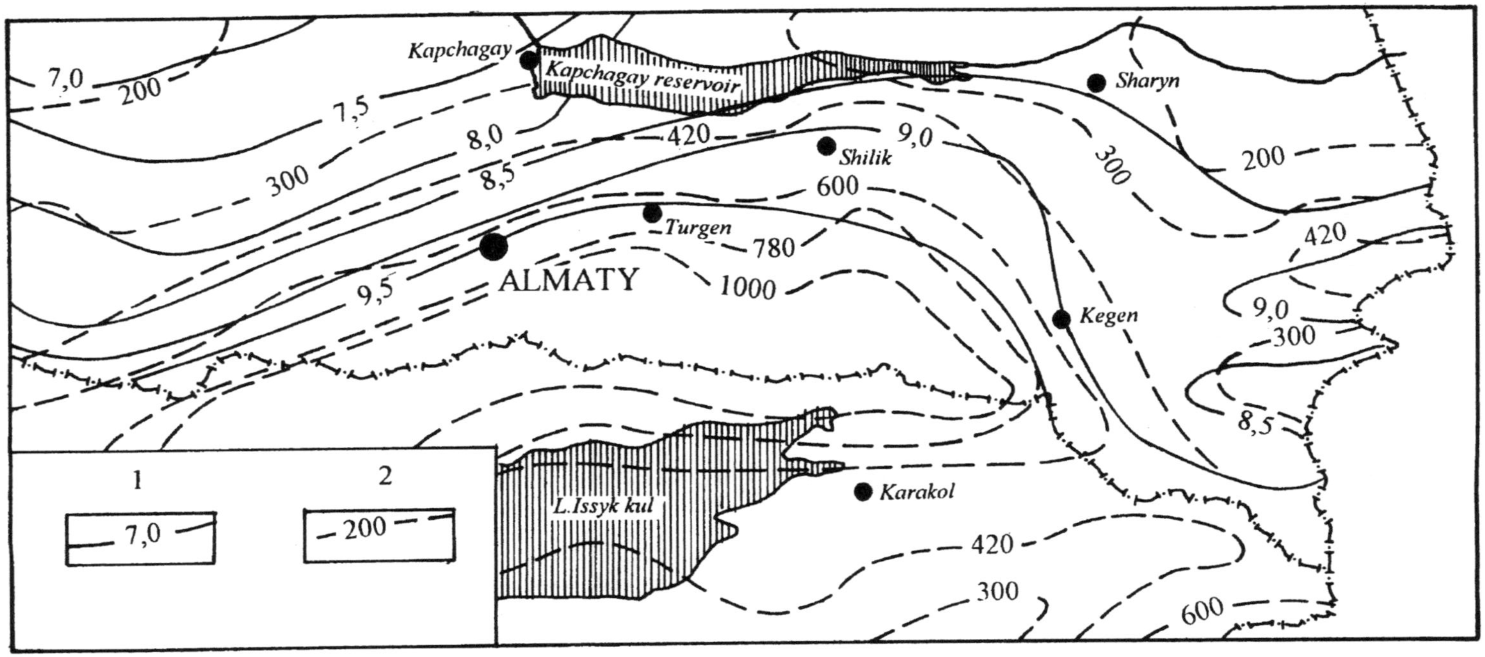

Figure 5. Map of detailed seismic zoning of Almaty industrial region. 1 = intensity of shock; 2 = isolines of maximum acceleration in cm/sec^2.

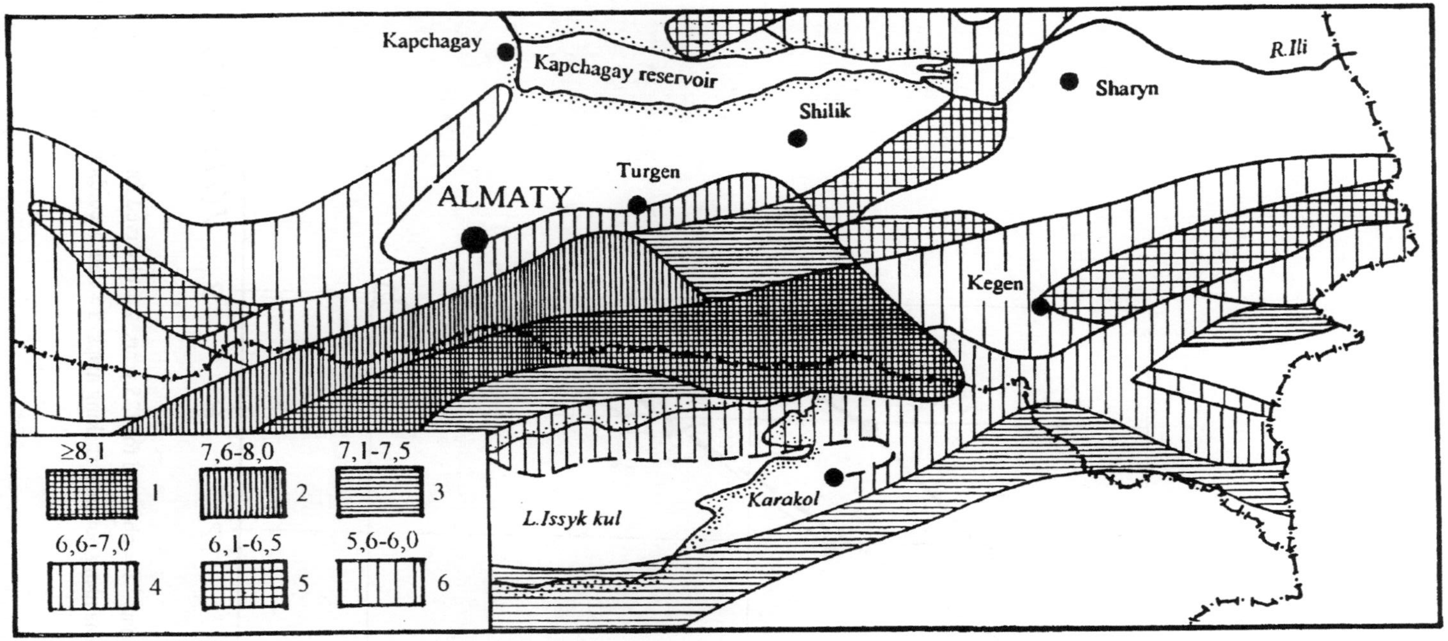

Figure 6. Map of seismic generating zones in Almaty industrial region. 1-6 = seismic generating zones and maximum size of expected earthquakes.

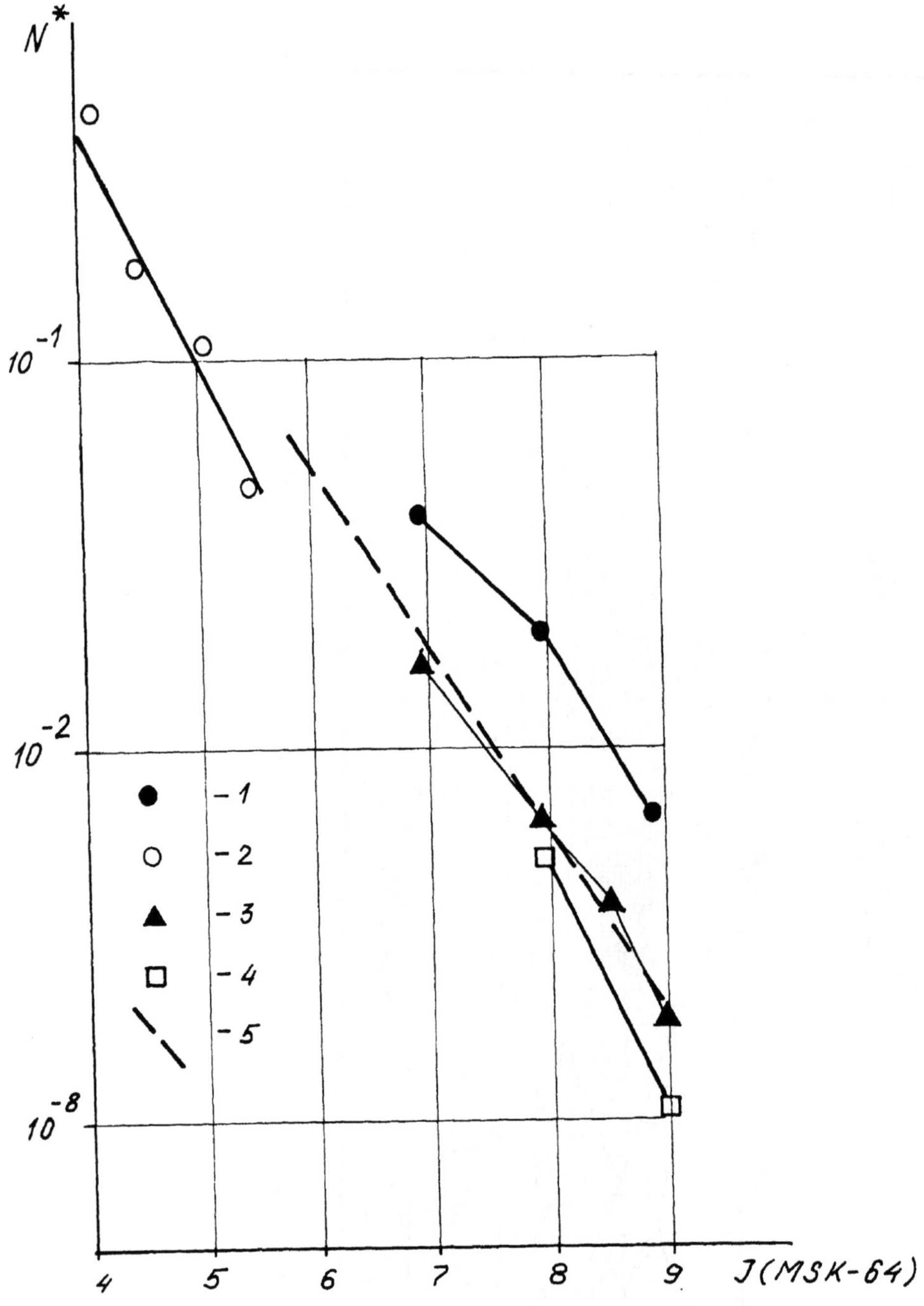

Figure 7. Intensity recurrence for Almaty. 1 = observed for 125 years; 2 = observed for 1928-1992; 3 = calculated according to Cornell; 4 = data using potential shaking maps; 5 = data according to average seismic regime.

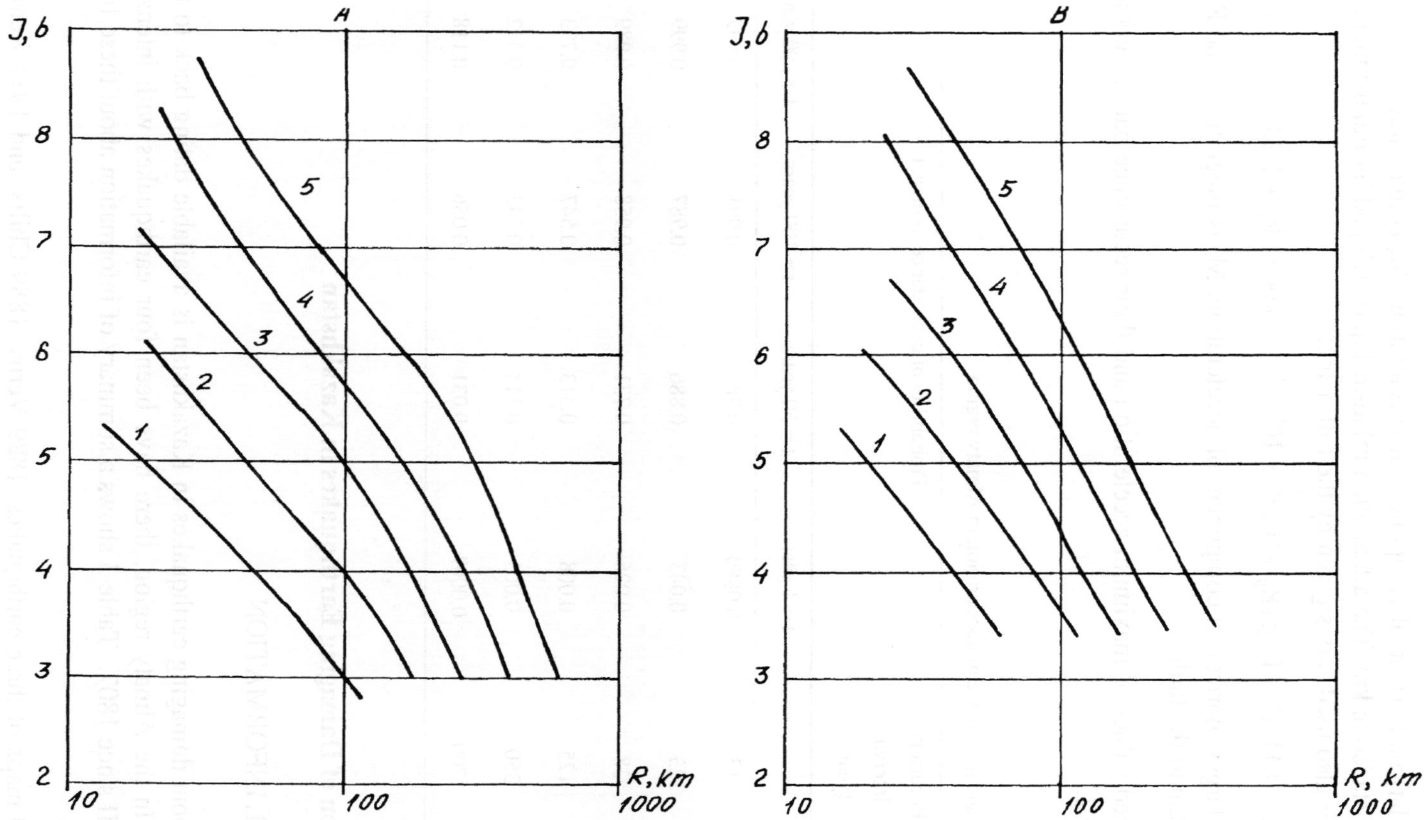

Figure 8. Graph of intensity attenuation for the Almaty region. A = along faults; B = across faults. 1 = M_{avg}: 4.5; 2 = M_{avg}: 5.5; 3 = M_{avg}: 6.5; 4 = M_{avg}: 7.5; 5 = M_{avg}: > 8.0.

$$\text{For } A \geq 160 \text{ cm/s}^2,\ \text{Log } A = 0.28\ M_s - 0.8 \text{ Log } R + 1.70 \qquad (3a)$$

$$\text{For } A < 160 \text{ cm/s}^2,\ \text{Log } A = 0.80\ M_s - 2.3 \text{ Log } R + 0.80 \qquad (3b)$$

where A is the largest horizontal component of acceleration, M_s is magnitude, and R is the epicentral distance in km The attenuation relationship developed by Fukushima and Tanaka (1990) is also used and is given by the following:

$$\text{Log } A = 0.41\ M_s - \text{Log } (R_p + 0.032 \text{ x } 10^{0.41\ Ms}) - 0.0034\ R_p + 1.23 \qquad (4)$$

where A is the largest horizontal component of acceleration, M_s is magnitude, and R is the distance in km to the fault.

The expected values of maximum acceleration and their recurrence parameters are shown in Table 2.

TABLE 2. Values and recurrence of acceleration in Almaty region

Acceleration (cm/s2)	Recurrence Interval (years)	Probability of exceedance in time, t			
		t = 1 year	t = 50 years	t = 100 years	t = 200 years
50	13	0.0880	0.986	0.991	0.999
100	23	0.043	0.886	0.987	0.999
200	38	0.026	0.731	0.958	0.989
400	125	0.008	0.313	0.547	0.793
600	350	0.003	0.114	0.243	0.412
800	1700	0.0007	0.031	0.056	0.118

3. Description of Damaging Earthquakes in Kazakhstan

3.1 GENERAL INFORMATION

Information about damaging earthquakes in Kazakhstan is available dating back to the early 1800s. In the Almaty region, there have been four earthquakes with intensity greater than VII since 1807. Table 3 shows a summary of information about these four earthquakes.

Isoseismal maps of three earthquakes, 1889 Verny, 1889 Chilik, and 1911 Kemin, are shown in Figures 9, 10, and 11, respectively.

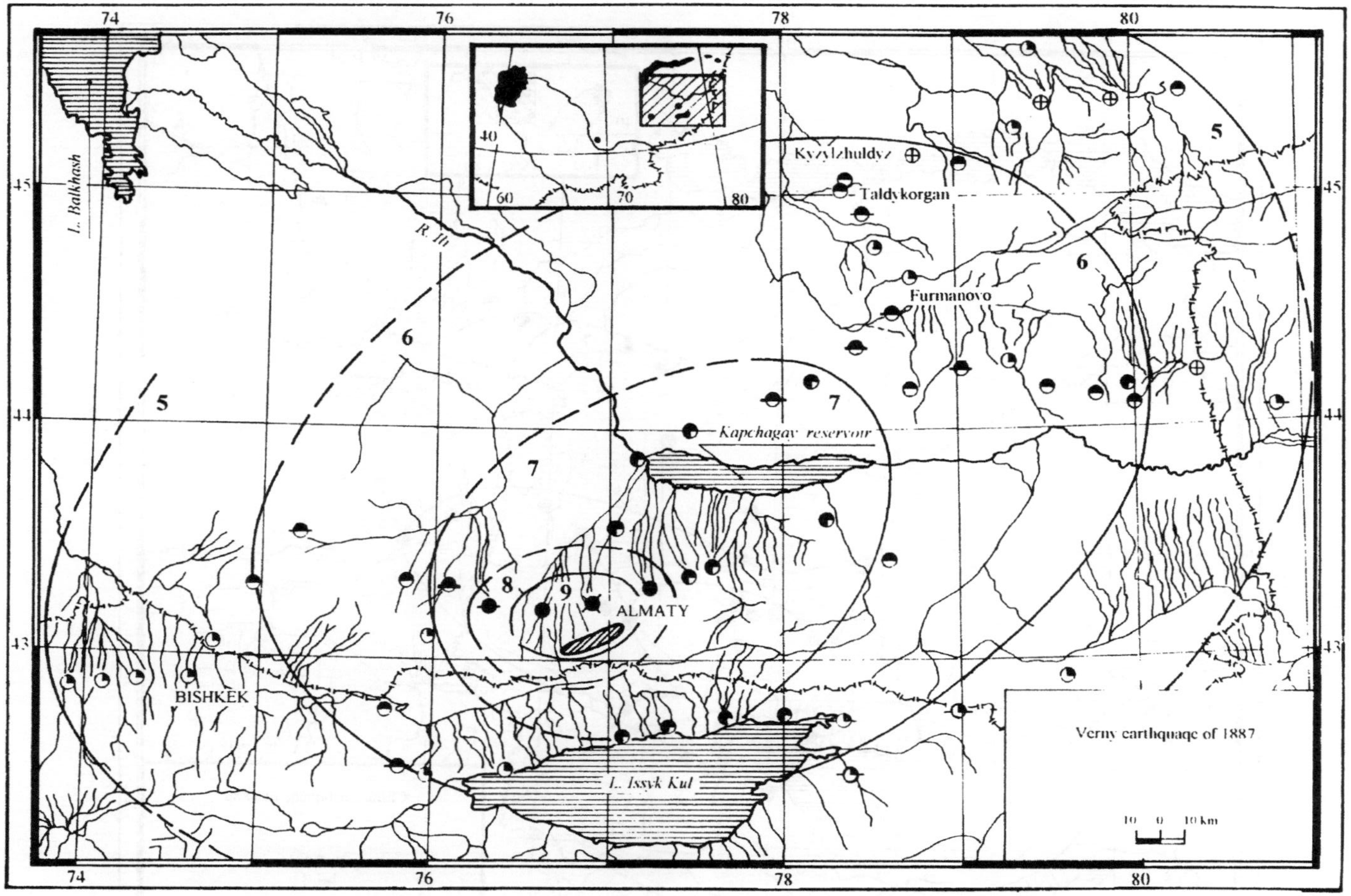

Figure 9. Isoseismal map of the M 7.3 Verny earthquake of 8/6/1887 (Sydkov, Nurmagambetov, Dosymov, Dzanuzakov, and Mushketov).

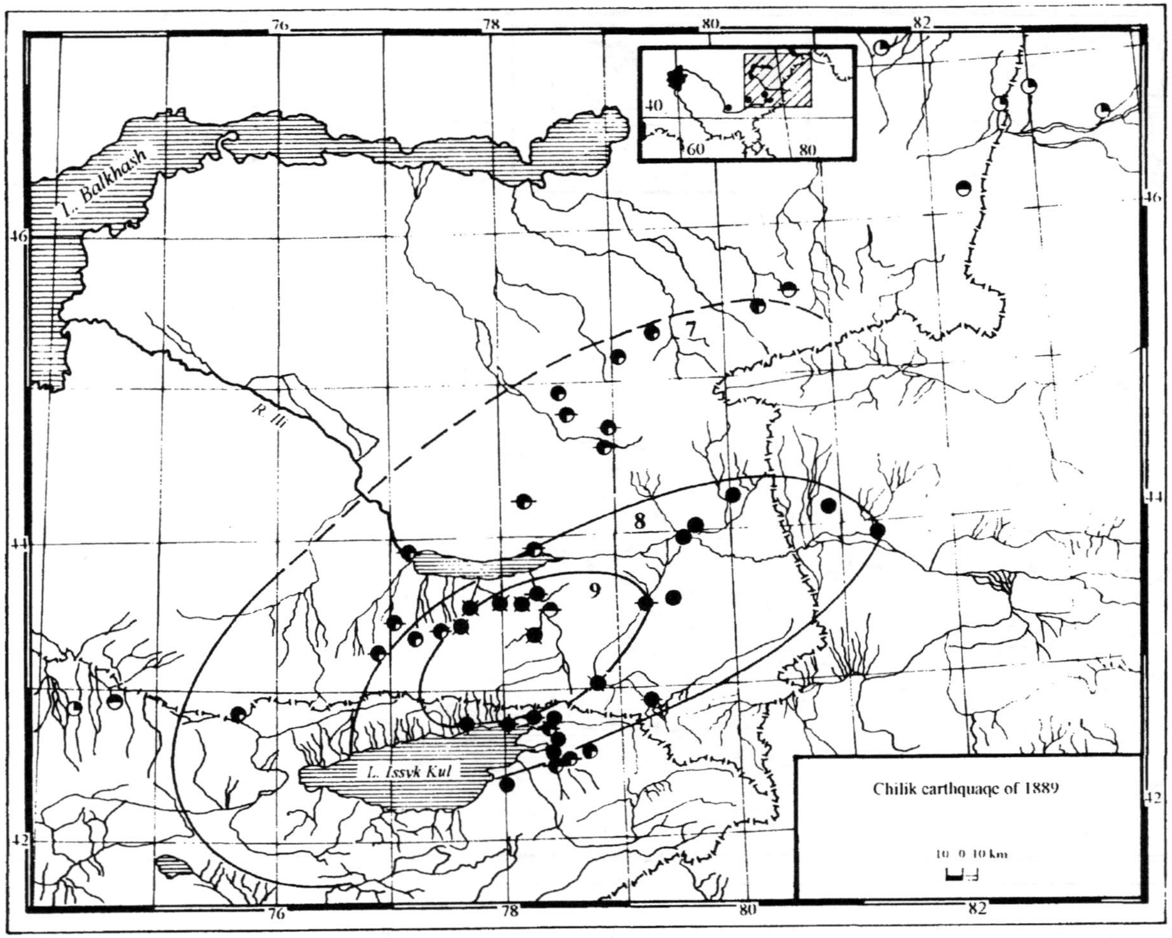

Figure 10. Isoseismal map of the M 8.4 Chilik earthquake of 11/7/1889 (Sydkov, Nurmagambetov, Dosymov, Dzanuzakov, Mushketov, Zenkov, and Aristova).

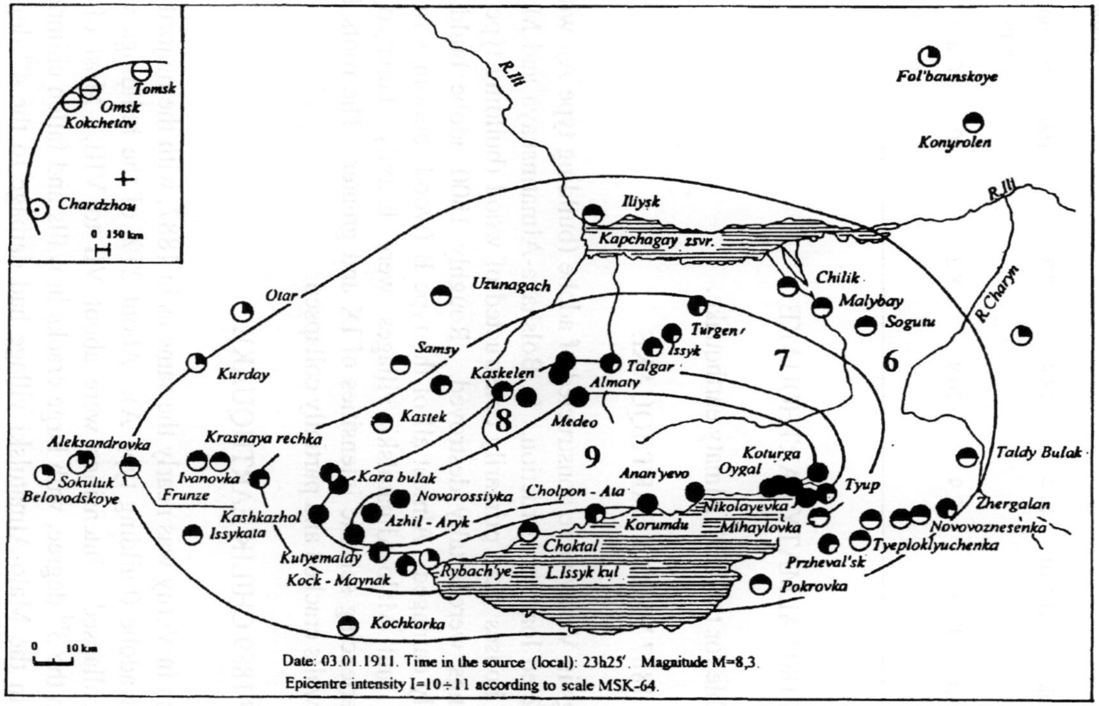

Figure 11. Isoseismal map of the Kemin earthquake of 1911.

TABLE 3. Summary of information about damaging earthquakes in Almaty

No.	Date	Name	Intensity	Coordinates		Magnitude	R (km)	Main Fault
				Lat (N)	Long (E)			
1	1807	Almaty	VIII-IX	43.1	76.9	6.7	15	Kemin-Ushkonur
2	June 8, 1887	Verny	IX-X	43.1	76.8	7.3	13	Kemin-Ushkonur
3	July 11, 1889	Chilik	VII-VIII	43.4	78.4	8.3	100	Baisorun-Chilik
4	Jan. 3, 1911	Kemin	IX-X	42.9	76.9	8.2	50	Chilik-Kemin

3.2 DESCRIPTION OF 1807 ALMATY EARTHQUAKE

No information is available for the 1807 Almaty earthquake.

3.3 DESCRIPTION OF 1887 VERNY EARTHQUAKE

The majority of buildings in Verny were constructed of adobe (building type A), with a few buildings of compacted loam construction. In Bolshaya-Almatinskaya and Malo-Almatulski villages, the houses were primarily constructed of wood (building type B). Type A (adobe) buildings were totally destroyed. Roughly 1800 adobe buildings collapsed or were heavily damaged. Almost all of the type B (wood) buildings in the Bolshaya-Almatinskaya and Malo-Almatulski villages were heavily damaged or partially collapsed, experiencing seismic intensities of IX and greater. The roofs fell, columns collapsed, and walls cracked and partially collapsed.

3.3 DESCRIPTION OF 1889 CHILIK EARTHQUAKE

The building construction in Verny was nearly the same as in 1887, with the majority of buildings constructed of adobe (building type A). About 27% of the buildings were heavily damaged and collapsed. Intensities were about VII to VIII. Most of the buildings had damage to the 3rd degree, with large cracks in walls and fallen chimneys. Some of the buildings in the Malo-Almatulski village had damage to the 4th degree, which includes collapse of inner walls and other building components.

3.4 DESCRIPTION OF 1911 KEMIN EARTHQUAKE

At the time of the earthquake, buildings were constructed of wood, adobe, mud, or fired bricks. The roofs were clay tile or thatch. Nearly 50% of the buildings were destroyed or needed major repairs. Damages occurred to the walls, roofs, stoves, and chimneys.

3.5 REASONS FOR EARTHQUAKE DAMAGE

Building damage can be attributed to the following factors: high intensity of shaking, proximity to earthquake epicenters, and cracks in building foundations. In addition, the quality of construction, such as adobe brick houses with thatch and clay roofs, and building foundations made of rubble, contributed to the damage.

3.6 SECONDARY EFFECTS

The earthquakes caused landslides, mudflows, surface heaves in loose and hard rocks, and surface soil cracks and heaves in the mountains of Zailiski and Kungey Alatau. Soil deformation was observed in several regions within Verny. Surface cracks as large as 1 meter wide and 5 meters deep were created by the 1911 Kemin earthquake.

4. Description of Building Construction in Kazakhstan

The description of building construction according to seismic resistance is defined according to "Methodical recommendations on introduction of passport system for the buildings of existing construction in Almaty and other settlements located in the seismic hazard regions of Kazakh SSR." These recommendations were developed in 1989 by the Institute KazNIISSA. In the recommendations, all buildings of existing construction are divided into 13 classes according to their seismic resistance.

Table 4 shows a summary of the number of buildings different levels of seismic vulnerability. In Table 4, seismic hazardous buildings refer primarily to brick buildings and individual houses built from local materials. Seismic resistant buildings refer to large panel buildings with monolithic reinforced concrete construction.

TABLE 4. Number, total area, and inhabitants of buildings with various levels of vulnerability

Seismic Hazardous Buildings			Buildings to be Examined			Seismic Resistant Buildings		
Total Number (x1000)	Total Area (x1000 sq. m.)	Number of Inhabitants	Total Number (x1000)	Total Area (x1000 sq. m.)	Number of Inhabitants	Total Number (x1000)	Total Area (x1000 sq. m.)	Number of Inhabitants
14.2	4,700	200,000	5.3	8,000	250,000	3.5	10,500	750,000

The most dangerous buildings in terms of seismic resistance qualities are the following:

- Buildings constructed of local materials (e.g., adobe or mud)
- Brick buildings constructed without seismic considerations or with insufficient lateral resistance
- Buildings with flexible ground floors
- Buildings in poor condition due to physical wear

5. Research Institutes and Organizations in Kazakhstan

The primary organization working on SNiP and regional standard documents for regulating seismic construction is the Kazakh Research and Design-Experimental Institute of Seismic Resistant Construction and Architecture (KazNIISSA). The address of KazNIISSA is 135 zh, Gagarin PR., Almaty, 480033.

Other companies and organizations using construction regulation documents in their project work include the following:

- Almaty Domostroitelny Kombinat (90, Satpayev Str.)
- Basiz Company
- Almatykultbytstroi (20, Dzhandosov Str.)
- Montazhspetsstroy (10a, Abay Pr.)
- Sredazenergostroy (66, Suyunbay Str.)

6. Regulations for Seismic Resistant Design and Construction in Kazakhstan

Construction norms and regulations that have been in force since the beginning of the 1960s include the following:

- SNiP II-A.12-62: "Construction in Seismic Regions. Design Norms." Approved by the resolution of the USSR Gosstroi. Valid from 1970 to 1980.
- SNiP II-7-81* (before 1990 SNiP II-7-81): "Construction in Seismic Regions. Design Norms." Approved by the resolution of the USSR Gosstroi. Valid from 1981 to the present.
- RSN 10-70: "Construction of Alma-Ata city and adjoining territories taking into account seismic micro zoning." Approved by the resolution of Kaz SSR Gosstroi. Valid from 1970 to 1983.

- RSN 10-83: "Construction of Alma-Ata city and adjoining territories taking into account seismic micro zoning." Approved by the resolution of Kaz SSR Gosstroi. Valid from 1983 to 1995.
- SN RK B2. 2-7-95: "Construction of Alma-Ata city and adjoining territories taking into account seismic micro zoning." Approved by the resolution of Minstroi of Kazakhstan. Valid from 1995 to the present.

6.1 THEORETICAL BASIS FOR REGULATIONS

The dynamic theory of defining design seismic loads was used in the development of the regulations SNIP II-A.12-62 and SNIP II-A.12-69. In the regulations, the definition of seismic loads on the building were based on seismic coefficients reflecting the intensity of expected earthquakes and on graphs of the dynamic coefficient, β(T), reflecting the dynamic effects of the seismic loads.

In SNiP II-7-81 the methods for defining design seismic loads were developed further. For its time, SNiP II-7-81 is a rather progressive regulatory document containing a number of new regulations for defining seismic loads for building design and construction. These regulations, without changing the essence of the dynamic method of seismic load definition reflected the current level of the theory and practice of seismic resistant construction. The features that are different in SNiP II-7-81 from the previous document are the following:

- Consideration of local soil conditions in seismic loads definition
- Definition of design values based on actual levels of accelerations of foundations during earthquakes
- Consideration of earthquakes recurrence

The influence of local soil conditions was addressed by introducing into the regulations of SNiP II-7-81 individual dynamic coefficient graphs which characterize the effect of seismic loads traces on surfaces with firm, moderately firm (moderately soft) and soft soils.

Elastic structural deformation is assumed when defining the seismic loads for building design. In SNiP II-7-81, it was suggested that actual instrumental recordings of earthquake shaking or synthesized accelerograms be used for defining the seismic load; however, the selection of accurate input motions was not well defined.

The design codes developed in Kazakhstan (RSN 10-70, RSN 10-83 and SN RK B.2 2-7-95 written by the institutes KazNIISSA and KazGIIZ) provided regulation for the construction of buildings and other facilities in Almaty and the adjoining territories. This includes both new design and rehabilitation of existing structures

Since 1981, the Almaty region has experienced several strong earthquakes. These events have provided specialists with information about building performance, especially the adequacy of the current design codes, and the regional socio-economic consequences of large earthquakes. In addition, the results from engineering analysis and the recorded ground motions showed the need for specifying the design seismic load level in the regulations.

A project to develop a new design code for construction of buildings in seismic regions of Kazakhstan was undertaken by specialists from KazNIISSA. In this project, the accumulated experimental data was combined with the theoretical basis of the SNiP II-7-81 document. In the new code, design seismic loads on buildings may be defined by the spectral method, by actual earthquake recordings, or by synthesized accelerograms.

Formulas for calculating the design seismic loads, S_{ik}, by the spectral method are as follows:

$$S_{ik} = K_1 K_2 K_3 S_{oik} \tag{5}$$

$$S_{oik} = Q_k A \beta_i K_\psi \eta_{ik} \tag{6}$$

where K_1 is a coefficient that accounts for the impact that the damage to the building would have ranging from 1 for ordinary residential buildings to 4 for important facilities, K_2 is a coefficient that accounts for the seismic resistance of the building ranging from 0.2 for large panel buildings to 0.35 for buildings with brick walls, and K_3 is a coefficient that accounts for the height of the building and is defined as follows:

$$K_3 = 1 + 0.06(n - 5) \tag{7}$$

where n is the number of stories. All other coefficients are the same as those described in the SNiP II-7-81 document. The basis for the values of coefficients K_1, K_2, and K_3 are the results of engineering analysis following strong earthquakes, results of dynamic testing of over 40 full scale and model buildings, and results of numerous tests of individual structural elements.

The dynamic coefficient in Equation 6, $\beta(T)$, is defined according to the graphs shown in Figure 12 and the values in Table 5. These relationships were developed using ground motion recordings of about 150 buildings along with information about the spectral content of the ground motions. It is expected that this definition of $\beta(T)$ describes the dynamic effects of seismic loads on buildings better than the SNiP II-7-81 document.

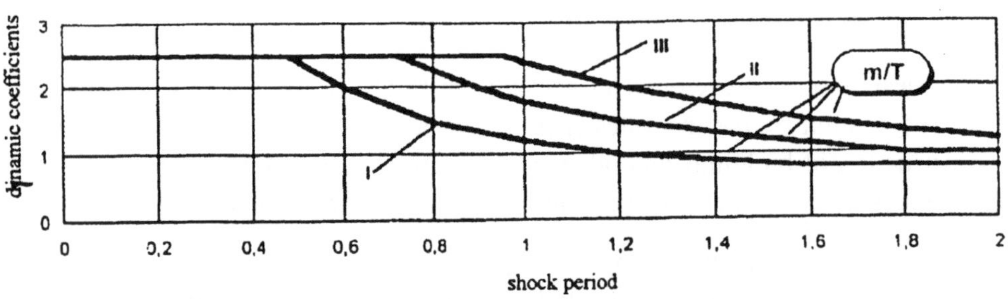

Figure 12. Graph of dynamic coefficient as a function of period in seconds.

TABLE 5. Parameters for defining dynamic coefficient, β(T), for three ground conditions

Ground category according to seismic properties	T_1	β_{max}	β_{min}
I	0.48	2.5	0.8
II	0.72	2.5	1.0
III	0.96	2.5	1.2

The concept of effective peak acceleration is used to define the coefficient A in Equation 6. Empirical data was used to define the level of acceleration corresponding to each level of intensity, as indicated in Table 6.

TABLE 6. Values of coefficient A corresponding to various levels of seismic intensity

Seismic Intensity	A
VII	0.125
VIII	0.25
IX	0.5
X	0.9

According to the new design regulations, dynamic calculations using recorded ground motions or synthesized accelerograms are required for the following buildings:

1. Important structures that should have little damage in a strong earthquake
2. Buildings taller than 50 meters with plan or vertical irregularities
3. Buildings with relatively new design features such as seismic isolation

For buildings of type 1 or 2 described above, linear elastic behavior is assumed for the translation of earthquake ground motions to the coefficients required in Equations 5, 6, and 7. For buildings of type 3 described above, design models must be based on nonlinear theory and experimental results.

The selected ground motions for the dynamic calculations should be the most typical for the region in terms of spectral content, load level, duration, and earthquake mechanisms. The 5% damped response spectra of the selected ground motion should correspond to that shown in Figure 13 and Table 7.

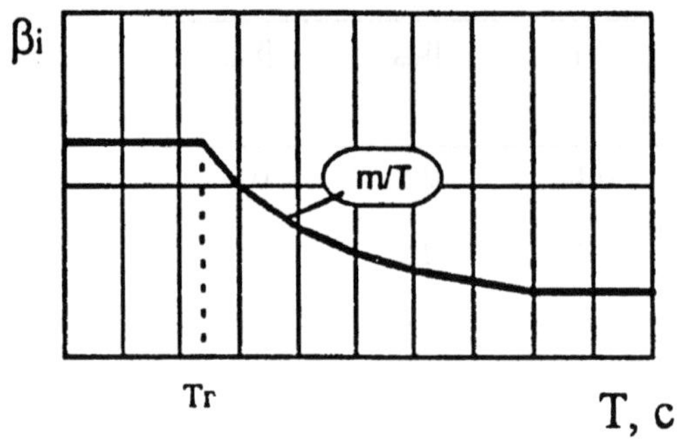

Figure 13. Graph of response spectrum specified in design regulations.

TABLE 7. Parameters of response spectra for dynamic calculations corresponding to three soil conditions

Ground category corresponding to seismic properties	T_1	T_2	$A\beta_{max}$	n	m
I	0.10	0.40	2.5A	15	1.0
II	0.15	0.60	2.5A	10	1.5
III	0.20	0.90	2.5A	7.5	2.25

It should be stressed that the requirements of the design regulations described here are considered the minimum necessary to ensure safety of buildings and occupants during earthquakes.

6.2 ENFORCEMENT OF DESIGN REGULATIONS

All project documents developed by design organizations in Kazakhstan must undergo review by the State Construction Expertise (Gosekspertisa) and the Ministry of Building, Dwelling and Construction of the territory of the republic of Kazakhstan.

7. Seismic Strengthening of Existing Buildings in Kazakhstan

7.1 METHODS FOR SEISMIC STRENGTHENING OF BUILDINGS

Seismic strengthening of buildings and other structures is regulated by the "Manual on Increasing Bearing Structures of Buildings and Constructions of Industrial Enterprises under Reconstruction, Situated in II and III Zones of Almaty." The manual was developed in 1986 by the institute KazNIISSA, with participation from the institute

TbilZNIIEP. The manual describes the methods for assessing the existing condition and necessary seismic strengthening of building undergoing retrofit.

Depending on the seismic capacity and existing condition of structural elements in the buildings, the manual recommends the following techniques be used when not changing the overall structural scheme of the original design:

- Wrap columns with reinforced concrete or steel rings
- Shotcrete existing walls
- Glue steel sheets or corrugated glass
- Inject polymer materials into cracks
- Add vertical reinforcement

When making changes to the overall structural scheme of the original design, the following techniques are recommended:

- Add steel or reinforced concrete vertical elements and unload some bearing elements
- Strengthen bearing elements, such as partitions, by increasing stiffness
- Increase stiffness between floors and between separate structural elements

7.2 ORGANIZATIONS WORKING ON SEISMIC STRENGTHENING

Design projects for strengthening and reconstruction of existing buildings may be carried out by all design organizations with participation, advice, and expertise provided by the institute KazNIISSA. These projects are regulated by the "Manual on Increasing Bearing Structures of Buildings and Constructions of Industrial Enterprises under Reconstruction, Situated in II and III Zones of Almaty," which was described in the previous section.

7.3 OBSTACLES TO SEISMIC STRENGTHENING

The main obstacle to seismic strengthening of existing buildings in Kazakhstan is the lack of available funding.

7.4 DESIGN LEVELS FOR SEISMIC STRENGTHENING

In general, the seismic strengthening of buildings is done to a level that corresponds to the expected seismic intensity at the building site.

8. New Approaches to Seismic Design in Kazakhstan

Seismic resistant design of new buildings is being carried out by all design organizations in Kazakhstan with assistance from KazNIISSA. Design is regulated by SNiP II-7-81*, SN RK B.2.2-7-95 ("Construction of Almaty and Adjoining Territories Taking into Account Seismic Micro-Zoning"), and other regulations that have been in force in the territories of the former USSR. Design levels correspond to the prescribed seismic load for the location in which the building will be constructed. In general, the design regulations are enforced for all new buildings, with the exception of individual residences in rural areas which are often built from local materials such as adobe and clay.

The following items will help to ensure the adequate seismic resistance of new buildings:

- Strict adherence to the design regulations
- Laboratory testing of proposed new designs
- Strict supervision of the construction process to ensure high quality
- Clear design guidelines in newly developed design documents

The primary obstacles to the proper design and construction of seismic resistant residential buildings include the following:

- Qualified specialists leaving the institutes
- Lack of available information about new structural design research and seismic resistant construction techniques
- Inexperienced new small design firms
- Untested new materials and techniques are appearing in the construction market

9. Optimal Methods for Defining Seismic Loads in Kazakhstan

The current design documents define seismic load in terms of intensity, the value of which can be found on seismic zoning maps for the region. However, response spectra and recorded ground motion time histories are the most useful data for defining seismic loads at a building site.

In Kazakhstan, the return period used in new design should be 1000 years for standard buildings and 10000 years for the most important buildings.

Seismic motion recording instruments are registered and maintained by the Institute of Seismology for instruments on the ground and the strong motion service of KazNIISSA in Almaty for instruments in buildings. The maximum acceleration is 0.65g on ground for intensity VIII and 0.05g in a building for intensity VI.

10. Scientific and Technical Cooperation for Reducing Seismic Risk

Scientific and technical collaboration should take place in engineering seismology, theoretical and experimental investigations of the seismic resistance of buildings, modern techniques of seismic hazard and seismic risk estimation, and earthquake preparedness of the residents.

Possible forms of collaboration include the following:

- A unified Central Asian information center that will collect, summarize, and distribute the results of research work conducted in various countries around the world
- Joint research projects on theoretical and experimental investigations into seismic resistant construction and seismology
- Annual workshops, conferences, and seminars, including discussion of specific topics
- Working group made up of members from various countries to develop new design regulations for construction in seismic regions
- Unified system for training scientists

10.1 PRIMARY FUNCTION OF A UNIFIED INFORMATION CENTER

The main functions of a unified Central Asian information center should include:

- Supplying Central Asians and others with results of scientific research in areas of mutual interest
- Fostering cooperative research work among specialists from various countries
- Funding joint research projects
- Conducting independent review of projects involving design and construction of new buildings in seismic regions

The center could be created within the institute KazNIISSA or be an independent organization.

10.2 ISSUES TO BE INCLUDED IN FUTURE WORKSHOPS

The following issues should be on the agenda for discussion at future workshops on seismic risk in Central Asia:

- Scales for assessment of earthquake intensity
- Seismic resistance of school buildings constructed from typical designs
- Reconstruction and strengthening of existing buildings

10.3 SPECIFIC ITEMS TO FOSTER INTERNATIONAL COLLABORATION

The following items are specific areas where international collaboration would be beneficial in reducing seismic risk in Central Asia:

- Creating doctoral programs or visiting scholar programs on issues of engineering seismology and seismic resistant design
- Acquiring strong ground motion recordings during large earthquakes
- Obtaining the latest publications and monographs on issues of seismic risk and reliability of buildings and other structures, and on defining seismic load levels for design
- Conducting reviews of modern design and construction of structures located in seismic regions
- Equipping strong ground motion recording stations in the region of Almaty with modern digital devices and computer programs for processing the recorded data
- Creating a database of strong motion recording data

SEISMIC HAZARD AND BUILDING VULNERABILITY IN KYRGYZSTAN

S. IMANBEKOV
Seismic-Resistant Construction Institute, Kyrgyzstan
K. DZHANUZAKOV
Institute of Seismology, Kyrgyzstan
S. URANOVA
Seismic-Resistant Construction Institute, Kyrgyzstan
A. FROLOVA
Institute of Seismology, Kyrgyzstan

1. Main Data on the Capital City of Kyrgyzstan

The capital city of Kyrgyzstan is Bishkek; the former names were Frunze and Pishkek. Bishkek is situated at the foot of the Kyrgyz ridge in the center of the Chu valley at an elevation of 750 meters above sea level. Bishkek lies in the plain formed by the joining of the river deposits of the Ala-Archa, Alamedin, and other rivers. The geographic coordinates of the city are approximately 42°49′ north latitude and 74°38′ east longitude. The city was founded in 1825, has a size of 17,318 hectares, and has a population of about 780,000 inhabitants.

The Chu depression, situated between the Chu-Ili mountains in the northeast and the Kyrgyz mountain ridge in the south, is a northerly sloping alluvial-proluvial plain, formed by deposits of the south tributary stream to the Chu river. The Chu depression has absolute elevation levels of 700 to 1000 meters. To the south of the city lies a latitudinally stretched zone of adirs, which are foothills that rise to an elevation of 2 kilometers in some locations. The north slope of the Kyrgyz range has a length of 30 to 40 km and a watershed elevation of 4985 m. The range has 3 layers of tectonic relief and was formed in the Oligocene period. The foothills are characterized by 2 layers of tectonic relief and were formed in the early Pleistocene period. The depression is the low layer of relief in which surfaces of later Pleistocene and Holocene terraces can be seen.

1.1 SHORT GEOMORPHOLOGICAL DESCRIPTION OF BISHKEK

There are three types of relief within the territory of Bishkek. They are as follows:

1. Lower mountain terrains of alluvial deposits from the Alamedin and Ala-Archa rivers (dry deltas)

S.A. King et al. (eds.),
Seismic Hazard and Building Vulnerability in Post-Soviet Central Asian Republics, 93–105.

2. Outlying portions of alluvial deposits of Alamedin and Ala-Archa rivers
3. Sloping alluvial-proluvial plains

A lower mountain terrain exists in the southern part of the city. It is bounded on the south by the foothills of the Kyrgyz range, and in the north it joins the outlying parts of the Alamedin and Ala-Archa rivers. The terrain stretches from east to west and the width of the zone within the city is 0.5 to 9.5 km. The surface of the lower mountain terrain changes in elevation from 980 meters in the south to 1730 meters in the north. A slightly folded surface in the mountain terrain is caused by the presence of deposits of the Alamedin and Ala-Archa rivers and dry deltas. The rivers cross the terrain from south to north. The terrain is characterized by crushed rocks covered by layers of clayey soils of 0.2 to 2.0 meters thick. Alluvial deposits are firm with a thickness reaching hundreds of meters in some locations.

The outer parts of deposits of the Alamedin and Ala-Archa rivers cover a width of 0.05 to 1.4 km. from east to west in the central part of the city. The surface changes in elevation from 760 to 725 meters. The top layer is clay that varies from 2.0 to 5.0 meters in thickness. Under the clay are alluvial deposits from 0.2 to 2.0 meters thick with layers and lenses of clayey soils.

1.2 SEISMOTECTONIC DESCRIPTION OF BISHKEK

The region of Bishkek is in the North Tien-Shan seismic zone. Two zones of seismicity exist, which conform to the structure of the Chu depression. They are the North Chu, which coincides with the foot of the Kindyktass range, and the South Chu, which coincides with the adirs range running latitudinally. The North Chu zone is represented by a transition of the south side of the Kindyktass range into the Chu monoclinal, which in a number of places includes a small number of steeply inclined faults. Earthquakes with magnitudes less than 7 and intensities reaching VIII have occurred in this zone. The Chu monoclinal stops suddenly in the south by a flexure faulting zone, south of which the lower Kyrgyz bending flexure has lowered the depression foundation up to 4.5 km. The south part of the flexure, beginning with the middle Pleistocene age, is under the influence of tectonic movement, and as a result, a foothill range with peaks up to 2 km has appeared. It is situated between the side faults of Issik-Atinsky in the north and Shamsi-Tyundyuksky in the south. Both faults have west-northwest stretching and decline into a body of depression. To the west of the Alamedin River, there are deposits that have been placed by the Chonkurak side fault, which further to the west delimits the Chu depression and its mountain framing.

Amplitudes of vertical displacement in the fault wings have been measured up to 6 km in the Bishkek region. In almost all areas the vertical displacements break not only Neogene deposits, but also Quaternary terraces and Holocene alluvial deposits. In addition, the side faults are often segmented by diagonal faults of northeast stretching (e.g., Aksu). In places of such crossings or specific junctions (e.g., Aksu and Shamsin), areas of paleoseismic displacements have been measured and earthquakes have occurred. Thus a Belovodsk earthquake is tied to the Aksu junction, and a Balasogun earthquake to the Shamsin junction.

The South Chu seismicity zone has as its eastern edge the Kemin-Chilik zone, which is associated with the Chu-Ili, Kebin, Chilik, and other earthquakes with magnitudes up to 8.3 and intensities up to XI. Analysis of Paleoseismic dislocations shows directly that disasters of this magnitude used to occur in this zone, as least from the middle Pleistocene age.

1.3 GENERAL SEISMIC ZONING FOR BISHKEK

The initial maximum intensity estimate for Bishkek is assumed to be IX, according to the map of seismic zoning of Kyrgyzstan created in 1995. The primary engineering geological factors affecting the seismic intensity are tectonic zones with thrusting faulting structure, lithological structure, thickness of clay deposits, depth of groundwater, water saturation of the soil, and density of clayey and sandy soils.

The seismic zoning according to soil conditions is made in accordance with Table 1 requirements described in SNiP-7-81 and RSCH-85. Based on an analysis of the engineering geological, geomorphological, and hydrogeological materials in the Bishkek territory with respect to seismic characteristics, zones of soil types I, II, and III were identified.

Medium soils are assumed to be soils with layers of loam, sandy loam, and sandy soils with seams of alluvial deposits, covered on the surface with loam and sandy loam up to 5 meters thick. The groundwater is located at a depth of 1 to 5 meters. The increase in seismic intensity for other soil conditions is determined relative to these medium soils. The initial maximum estimate of intensity of IX for Bishkek is assumed to be for medium soils, for which the value of intensity increase is zero.

Zones with intensity VIII, and category I soils according to seismic properties, occupy the area in the south part of the city, mainly in the limits of the lower mountain terrain. The soils here are represented by layers of large disintegrated rock formations with insignificant cover layers of loamy soil up to 2 meters thick. The level of groundwater here is at a depth of 10 meters.

Zones with intensity IX, and category II soils according to seismic properties, occupy the area to the north of the intensity VIII zones described above, and refer to the gently sloping alluvial-proluvial plains. The soils are formed by layers of loamy soils, sandy loam, and sands with alluvial seams. Small local areas of category II soils are found in the zone of category III soils. The depth of groundwater is from 4 to 10 meters.

Zones with intensity greater than IX, and category III soils according to seismic properties, occupy all the northern part of the Bishkek territory, including the gently sloping plain with a flat surface. The lithological structure of these zones includes layers of loam, sandy loam, clay, sand with seams of silt, and peat. The level of groundwater is at a depth of about 5 meters or more.

2. Characteristics of Seismic Hazard and Expected Intensity in Kyrgyzstan

The region of the capital city of Bishkek in the Chu depression and its mountain frame refer to the North Tien-Shan seismic zone. The North Chu zone frames the Chu valley and is represented by the Kindyktass and Kastek subzones. The Kindyktass subzone has a maximum expected magnitude of 6.5. The recurrence interval for a magnitude 6.5 event is about 30 years, and the prevailing type of ground movement is upthrust. The Kastek subzone has a maximum magnitude of 7.0. The recurrence interval for a magnitude 7.0 event is about 100 years, and the prevailing type of ground movement is also upthrust.

The South Chu zone also contains many subzones, which are delineated according to structural and seismotectonic features. The Aksu-Shamshin subzone has a maximum magnitude of 7.5, a 300-600 year recurrence interval for a magnitude 7.5 event, and both upthrust and horizontal faulting movement. The Aspar-Kindyn subzone has a maximum magnitude of 7.0, a 100 year recurrence interval for a magnitude 7.0 event, and upthrust faulting movement. The Issik-Atinskaya subzone has a maximum magnitude of 6.5, a 30 year recurrence interval for a magnitude 6.5 event, and upthrust faulting movement. The Privodorazdel subzone has a maximum magnitude of 6.5, a 30 year recurrence interval for a magnitude 6.5 event, and both upthrust and horizontal faulting movement. The Belovodsko-Georgiyevskaya subzone has a maximum magnitude of 6.0 and both upthrust and horizontal faulting movement. The Chilik-Kemin subzone has a maximum magnitude of 8.0, a 1500-2000 year recurrence interval for a magnitude 8.0 event, and upthrust faulting movement. The West Kungey subzone has a maximum magnitude of 7.0 and a 100 year recurrence interval for a magnitude 7.0 event.

Isolines of earthquake recurrence intervals, T_y, with intensity of VIII to IX generally follow the configuration of the seismic zones. The recurrence intervals for events in the South Chu zone are 500, 200, and 100 years for earthquakes with intensity IX, VIII, and V, respectively.

Processing and analysis of macroseismic data for strong earthquakes have helped to define coefficients for seismic intensity attenuation. For most earthquakes, intensity attenuation coefficients are measured from 3.0 to 5.0. The average coefficient for the Bishkek region is 4.0, with 3.5 for areas along the structures and 4.5 for areas across the structures. Empirical formulas, showing relationships between intensity (I), magnitude (M), distance (r), and depth (h) of the earthquake, were obtained based on macroseismic data and are as follows:

$$I = 1.5\,M - 3.8\,\log(r^2 + h^2) + 3.6 \qquad \text{for medium radius} \qquad (1)$$

$$I = 1.5\,M - 3.4\,\log(r^2 + h^2) + 3.3 \qquad \text{for along structures} \qquad (2)$$

$$I = 1.5\,M - 4.4 \log (r^2 + h^2) + 4.2 \qquad \text{for across structures} \qquad (3)$$

The attenuation law for peak horizontal acceleration in the North Tien-Shan region developed by Aptikayev (1979) is as follows:

$$\text{Log } x = 0.2\,M - 0.8 \log R + 1.7 + C_1 \qquad \text{for } x \geq 160 \text{ cm/sec}^2 \qquad (4)$$

$$\text{Log } x = 0.8\,M - 2.3 \log R + 0.8 + C_2 \qquad \text{for } 10 \leq x \leq 160 \text{ cm/sec}^2 \qquad (5)$$

where x is the maximum amplitude of horizontal surface acceleration, M is the magnitude of the earthquake, R is the distance to the epicenter, C_1 is a constant that accounts for the type of earthquake motion, and C_2 is a constant that accounts for the type of soil. C_1 is equal to 0.2 for thrust faulting, 0.1 for thrust/strike-slip faulting, 0 for strike-slip faulting, -0.1 for normal/strike-slip faulting., and –0.2 for normal faulting. C_2 is equal to -0.15, 0, and 0.15 for soils of category I, II, and III, respectively.

Table 1 shows the values of maximum acceleration and predominant period in Bishkek for various seismic zones, assuming a minimum distance from the city to the zone. The values of acceleration and predominant period refer to the area of Bishkek that is characterized by soils of category II.

TABLE 1. Maximum values of acceleration and predominant period in Bishkek

Zone Number	Zone Name	M_{max}	Distance (km)	Predominant Period (sec)	Maximum Acceleration (cm/sec^2)
I	North Chu	6.5	26	0.16	280
IIa	Aksu-Shamsi	7.5	24	0.3	600
IIb	Issik-Aty	6.5	21	0.16	360
IIc	Privodorazdel	6.5	34	0.16	230
IId	Belovodsk-Georgiyevskaya	6.0	23	0.23	230
III	Kemin	8.0	99	0.40	280
IV	West Kungey	7.0	92	0.25	90

TABLE 2. Summary information on residential buildings in Bishkek

Type of Building	1962-1969 SNiP II-A-12-62			1970-1980 SNiP II-A-12-69			1981-present SNiP II-7-81		
	Area (x1000 m^2)	Number of Buildings	Number of Inhabitants	Area (x1000 m^2)	Number of Buildings	Number of Inhabitants	Area (x1000 m^2)	Number of Buildings	Number of Inhabitants
Brick buildings with longitudinal bearing walls	250	200	16000	250	200	16000	300	220	20000
Brick buildings with lateral bearing walls	200	130	13000	200	130	13000	120	80	8000
Large panel buildings with fixed steel joints	200	90	13000	-	-	-	-	-	-
Large panel buildings with concrete joints	60	30	4000	1200	400	80000	1740	500	116000
Buildings with walls of unreinforced blocks	126	100	8400	-	-	-	-	-	-
Frame buildings	-	-	-	20	10	1500	25	12	1800
Frame buildings with brick infill	-	-	-	20	15	1500	40	30	3000

3. Description of Damaging Earthquakes in Kyrgyzstan

3.1 GENERAL INFORMATION

Information about damaging earthquakes in Kyrgyzstan according to macroseismic data is available for the past 500 (± 100) years. During this period, about 10 strong earthquake have occurred in the territory of Bishkek.

The regional catalog of strong earthquakes includes the events that have occurred from the historical times up to the present day. The definitions of the main parameters of the strong earthquakes before the instrumental period (about 1929) were made only by analyzing the microseismic data. A summary isoseismal map of strong earthquakes was made by joining corresponding isoseismal maps of individual earthquakes. The area of intensity IX shaking makes up about 20% of the whole territory, the intensity VIII area covers about 30% of the region, and the area of intensity VII shaking corresponds to the remaining 50% of the territory of Bishkek.

3.2 DESCRIPTION OF EARTHQUAKE DAMAGE

The more recent earthquakes (1992 M 7.7 Suusamir, 1990 M 6.5 Baisoorun, and 1992 M 6.3 Kochkor-Aty) were felt in Bishkek with intensities of VI or less and very little damage occurred. The residential buildings in Bishkek during strong earthquakes (year 1911) included one-story houses with walls of clay materials (about 90%) and houses with wooden walls (about 10%). About 95% of the one-story houses with clay walls collapsed. These houses had no wooden frame and almost no seismic resistance.

4. Description of Building Construction in Kyrgyzstan

Table 2 shows a summary of the number of buildings and inhabitants of various construction types in the city of Bishkek, including the code under which they were designed. Note that not included in this table are about 380,000 people who live in houses that are individual buildings of one and two stories, 95% of which have not been designed to resist seismic forces.

The most dangerous buildings (among buildings of at least 3 stories) in terms of seismic resistance qualities are the following:

1. Buildings constructed of unreinforced blocks: These blocks are typically hollow with very small end surfaces that do not provide adequate adhesion. Strain is not properly carried due to the lack of adhesion. Elements of seismic strengthening do not conform to current design standards. There are about 100 building of this type in Bishkek, occupied by about 8500 people.
2. Brick buildings with exterior longitudinal bearing walls and without interior bearing walls: Post-earthquake analysis of these buildings shows low seismic

resistance when there are no interior longitudinal walls. Most of these buildings were constructed from 1960-1974. There are about 100 buildings of this type in Bishkek, occupied by about 8500 people.

3. Buildings with brick walls of low quality brick or without adequate seismic design considerations: Post-earthquake investigations have shown that the level of seismic vulnerability of 85% of these buildings is about 2 times as high as that specified in the current design codes. This is generally caused by the low quality of building construction and the lack of seismic considerations in the building design. About 70% of these buildings do not meet the current design code requirements. There are about 650 buildings of this type in Bishkek, occupied by about 70,000 people.

4. Large panel buildings of 464 series with fixed joint details: These buildings have low seismic resistance according to the current design standards. There are about 90 buildings of this type in Bishkek, occupied by about 23,000 people.

5. Research Institutes and Organizations in Kyrgyzstan

The standard documents regulating seismic design measures are being developed by the Head Institute on Seismic Resistant Construction of Kyrgyzstan, Kyrgyz Scientific Research and Designing Institute of Construction and Minarkhstroy (Ministry of Architecture and Construction). Their address is Kyrgyz NIIPStroitelstva, No. 2 Cholpon – Atinskaya St., VPZ, 720571, Bishkek.

The names and addresses of the major design organizations include the following:

- JSC Bishkekproyekt, 164 a, Chu prosp., 720001, Bishkek
- JSC Kyrgyzpromstroy, 4, Manas prosp., 720571, Bishkek
- JSC Kyrgyzpromproyect, 219, Chu prosp., 720571, Bishkek

The names and addresses of the main building organizations that are involved in mass residential building construction include the following:

- Corporation Azat, ½ Auezov St., 720571, Bishkek
- JSC PSF Bishkekkurulush, 12, Tolstoy St., 720571, Bishkek
- Stock Company Ailkurulush, 2a, Tistinov St., 720571, Bishkek

6. Regulations for Seismic Resistant Design and Construction in Kyrgyzstan

Building standards and regulations that have been in force since the beginning of the 1960s include the following:

- SNiP II-A, 12-62: "Construction in Seismic Regions, Design Norms;" approved by the resolution of the USSR Gosstroi; valid from 1970 to 1980

- SNiP II-7-81* (before 1990 SNiP II-7-81): "Construction in Seismic Regions, Design Norms;" approved by the resolution of the USSR Gosstroi; valid from 1981 to the present
- SNiP 2.01.02-94 KR: "Building in the zones of Kyrgyzstan with seismicity greater than intensity IX;" valid from 1994 to the present
- SNiP 2.01-93 KR: "Building of Bishkek territory taking into account seismic zoning and soil-geological conditions;" valid from 1993 to the present
- SNiP 31-01-95 KR: "Retrofitting of residential buildings of present building;" valid from 1995 to the present

6.1 THEORETICAL BASIS FOR REGULATIONS

In design regulations SNIP II-A. 12-62 and SNiP II-A. 12-69, maximum amplitudes of accelerations of soil vibration for a given intensity and general spectral content (in terms of the descending part of the graph of the dynamic coefficient) are used.

In SNiP II-7-81, in addition to other parameters, recurrence of earthquakes of various intensity levels, and distinctions of spectral structure and vibration intensity for different soil conditions are taken into account.

In SNiP 2.01.02-94 KR and SNiP 2.01.01-93 KR, regional variations in seismicity are taken into account for the definition of seismic loads. The recurrence of earthquake events with given amplitudes, duration of shaking, and predominant periods of motion are also included in these design documents. The method of seismic load definition is based on the theory of reliability and seismic risk.

The definition of seismic loading in SNiP 31-01-95 KR is made in accordance with SNiP II-7-81*, SNiP 2.01.02-94 KR, and SNiP 2.01.01-93 KR. All of the relevant parameters in the design documents are the same; however, different definitions are used to account for the intensity of the design earthquake and the location of the building.

6.2 ENFORCEMENT OF DESIGN REGULATIONS

All of the design documents concerning building and architecture in Kyrgyzstan undergo a review by experts from the agencies of State Expertise of the Ministry of Architecture and Construction of Kyrgyzstan. Checks are made to ensure the correspondence of the design and construction documents to the current standards.

7. Seismic Strengthening of Existing Buildings in Kyrgyzstan

7.1 METHODS FOR SEISMIC STRENGTHENING OF BUILDINGS

The following two methods are used to increase the seismic resistance of residential buildings in Kyrgyzstan:

- Addition of reinforced concrete layers or rings around existing elements
- Addition of steel members

7.2 ORGANIZATIONS WORKING ON SEISMIC STRENGTHENING

Organizations developing design documentation on reconstruction of existing buildings include: KyrgyzNIICtroitelstva, Bishkekproyekt, Kyrgyzprostroy, Kyrgyzpromproyekt, and Bishkekkurulush.

7.3 OBSTACLES TO SEISMIC STRENGTHENING

The main obstacles to seismic strengthening of existing buildings in Kyrgyzstan include the following:

- Lack of financing
- Lack of special standardized approved documentation
- Inadequate study as to how well the reinforcing elements work with the existing elements
- Lack of availability of adequate construction materials
- Inadequate experience of design organizations in the field of building strengthening and reconstruction

The inadequate seismic resistance of many buildings that have been strengthened is due to the lack of understanding of how the reinforcing elements will actually work with the existing elements, and poor construction quality.

7.4 DESIGN LEVELS FOR SEISMIC STRENGTHENING

In general, the seismic strengthening of buildings is done to a level that corresponds to the expected seismic intensity at the building site.

8. New Approaches to Seismic Design in Kyrgyzstan

The organizations KyrgyzNIICtroitelstva, Bishkekproyekt, and Kyrgyzprostroy are engaged in projects concerning the design of seismically resistant residential buildings.

Currently in Kyrgyzstan, design projects are typically based on recommendations provided in SNiP II-7-81*, SNiP 2.01.01-93 KR, and SNiP 2.01.02-94 KR. Buildings are designed to resist the seismic loading that is prescribed in the regulations. Permission to construct the building is given after an examination of the project by experts to assess its conformance to the standard requirements.

Buildings with different designs, which have all been designed in accordance with the current standards, have different degrees of reliability. To increase the reliability of the buildings at the design stage, it is necessary to make improvements to the design standards. The low level of seismic resistance of buildings that are designed to the current standards is typically caused by the following:

- Not following the current design standards
- Lack of development of new technical solutions for seismic resistant construction
- Break-up of the project teams because qualified personnel are leaving the organizations

9. Optimal Methods for Defining Seismic Loads in Kyrgyzstan

Methods for defining the seismic loading should be made clear to all designers, and take into consideration all the possible information, such as maximum amplitude of acceleration, spectral content, duration of shaking, and recurrence of events with given parameters. Development of seismic load definition methods should be done at research institutes on seismic resistant building in collaboration with seismology institutes.

Seismic loads should be based on seismic risk and reliability assessment and include the features of the seismic zone. In addition, comparisons should be made between the possible damage, considering the hazard at the building site, and the initial construction costs.

A service for strong motion recording exists in Kyrgyzstan. The maximum acceleration value that has been measured is 52 cm/sec^2, corresponding to intensity level VI.

10. Scientific and Technical Cooperation for Reducing Seismic Risk

It is necessary to have collaboration in the areas of seismology, seismic resistant theory, and scientific research on seismic resistant design and construction. Possible forms of collaboration include the following:

- Creating a unified system for training of scientific personnel

- Creating a unified information center, containing information about scientific research work done in the different republics, which would also collect and disseminate scientific and technical information on seismic resistant design and construction from around the world
- Conducting meetings and conferences
- Carrying out mutual research work and exchanging research results
- Obtaining consultation on design and construction methods

10.1 PRIMARY FUNCTION OF A UNIFIED INFORMATION CENTER

The need for a constant long-term coordinating center is evident. Without such a center, collaborative activities cannot be effective. The main purpose of such a center would be to work out the problems associated with coordinating collaboration in scientific research work between the fields of seismology and seismic resistant design and construction. The center would be a link between the Central Asian republics and the other countries of the world.

The permanent staff of the coordinating center should include two coordinators, a scientific secretary, an organizational secretary, and representatives from each republic.

10.2 ISSUES TO BE INCLUDED IN FUTURE WORKSHOPS

The following issues should be on the agenda for discussion at future workshops on seismic risk in Central Asia:

- Possible ways for increasing seismic resistance of school buildings
- Issues concerning reconstruction and reinforcement of existing buildings
- Development of scientific/technical policies for building design standards in seismic zones

10.3 SPECIFIC ITEMS TO FOSTER INTERNATIONAL COLLABORATION

The following items are specific areas where international collaboration would be beneficial in reducing seismic risk in Central Asia:

- Proposals on training in the United States, in particular special doctoral training in the areas of engineering seismology, seismic risk, seismic reliability of buildings, construction for seismic loads, and development of building design standards for seismic resistant construction
- Databases of free field and building ground motion recordings from strong earthquakes
- Programs for creating databases of existing buildings

- Collection of information from the United States and other countries that relates to seismic resistance of buildings, standards for seismic resistant design and construction, seismic risk and reliability of buildings, and seismic load specification
- Publication of results of research on topics of seismic construction and engineering seismology
- Supplying seismic stations with new devices for strong motion recordings
- Consultation on optimal methods for calculating response spectra from digital recordings of ground motion velocity and for calculating seismic risk for the specific conditions of the Central Asian republics

SEISMIC HAZARD AND BUILDING VULNERABILITY IN TAJIKISTAN

S. NEGMATULLAEV
Institute of Seismology and Seismic-Resistant Construction, Tajikistan
A. ISCUK
Institute of Seismology and Seismic-Resistant Construction, Tajikistan
Y. POTEKHIN
Tajik State Committee of Construction and Architecture, Tajikistan

1. Main Data on the Capital City of Tajikistan

The capital city of Tajikistan, Dushanbe, was founded in 1924. It was renamed to Stalinabad in 1925, and then back to Dushanbe in 1961. The city is situated at an elevation of 760 to 800 meters above sea level. The geographic coordinates are 38°34′ north latitude and 68°48′ east longitude.

1.1 SHORT GEOMORPHOLOGICAL DESCRIPTION OF DUSHANBE

The city of Dushanbe is situated on an alluvial-proluvial complex of sediments and is divided into two large regions. The left bank is the second layer over a flood-lands terrace of Dushanbe complex, overlaid by loess soils of sinking deformation with a thickness of 10 to 40 meters. The right bank contains crushed gravel-pebble deposits with a thickness of 30 to 80 meters on flood-lands. The first layer of the flood-lands terrace is comprised of proluvial deposits of loess loamy soils up to 200 meters thick.

1.2 SEISMOTECTONIC DESCRIPTION OF DUSHANBE

Dushanbe is situated on the boundary of two large geologic structures, the Hercynian structure of the South Tien-Shan and the Alpine complex of the Tajik Depression. The boundary region is the Gissar Valley, which is bounded on the north by the zone of the Gissar-Kokshaal fault and on the south by the zone of the Illyak fault. These faults are seismically active. Maximum magnitudes are 7.5 for the Gissar-Kokshaal fault and 6.5 for the Illyak fault. Dushanbe is located about 5 km from the Gissar-Kokshaal fault and about 25 km from the Illyak fault.

The maximum expected intensity for Dushanbe is IX with a recurrence rate of 1 in 1000 years according to the map of General Seismic Zoning – 1978 (GSZ-78) shown in Figure 1. In addition to Dushanbe, there are two large seismic zones as shown in Figure

S.A. King et al. (eds.),
Seismic Hazard and Building Vulnerability in Post-Soviet Central Asian Republics, 107–125.

2, South-Gissar with a possible maximum magnitude of 7.9 in the north, and Illyak-Vakhsh with a possible maximum magnitude of 6.5 in the south.

1.3 GENERAL SEISMIC ZONING FOR DUSHANBE

According to the GSZ-78 map, the Dushanbe territory has en expected seismic intensity of IX.

1.4 TOTAL AREA AND POPULATION OF DUSHANBE

The total area of Dushanbe is approximately 140 square miles. The population in 1987 was about 570,000, but now the population of the city is about 1,100,000.

1.5 DESCRIPTION OF SOIL CONDITIONS IN DUSHANBE

The soil conditions are varied throughout Dushanbe, but in general they are of two types of deposits, loess loamy deposits and gravel-pebble deposits. The soil conditions are complicated by layers of buried soils and lenses of subsurface water. Because of the potential for loess loamy soils to experience subsidence deformation, seismic resistant building design and construction are complicated. Many buildings (about 40%) located on the left bank of Dushanbe are in danger of experiencing damage due to subsidence deformation.

2. Characteristics of Seismic Hazard and Expected Intensity in Tajikistan

According to the map of seismic zoning of Dushanbe that was made in 1975, two areas within the city have expected seismic intensity levels of VIII and IX. Both deep focus and shallow earthquakes are possible causing varied frequencies of earthquake motion. For this reason, zoning was made for two types of buildings, less than 5 floors which are likely to be affected by local shallow earthquakes, and 5 or more floors which are likely to be affected by long period motions of long duration from deep focus Pamir-Hindukush earthquakes.

Figures 1, 2, and 3 show the General Seismic Zoning (GSZ) map, the latest map of possible earthquake epicenters (1990), and the map of seismic microzoning in the territory of Dushanbe, respectively. As shown in Figure 1, the expected seismic intensity for the territory of Dushanbe is IX with a recurrence rate of 1 in 1000 years.

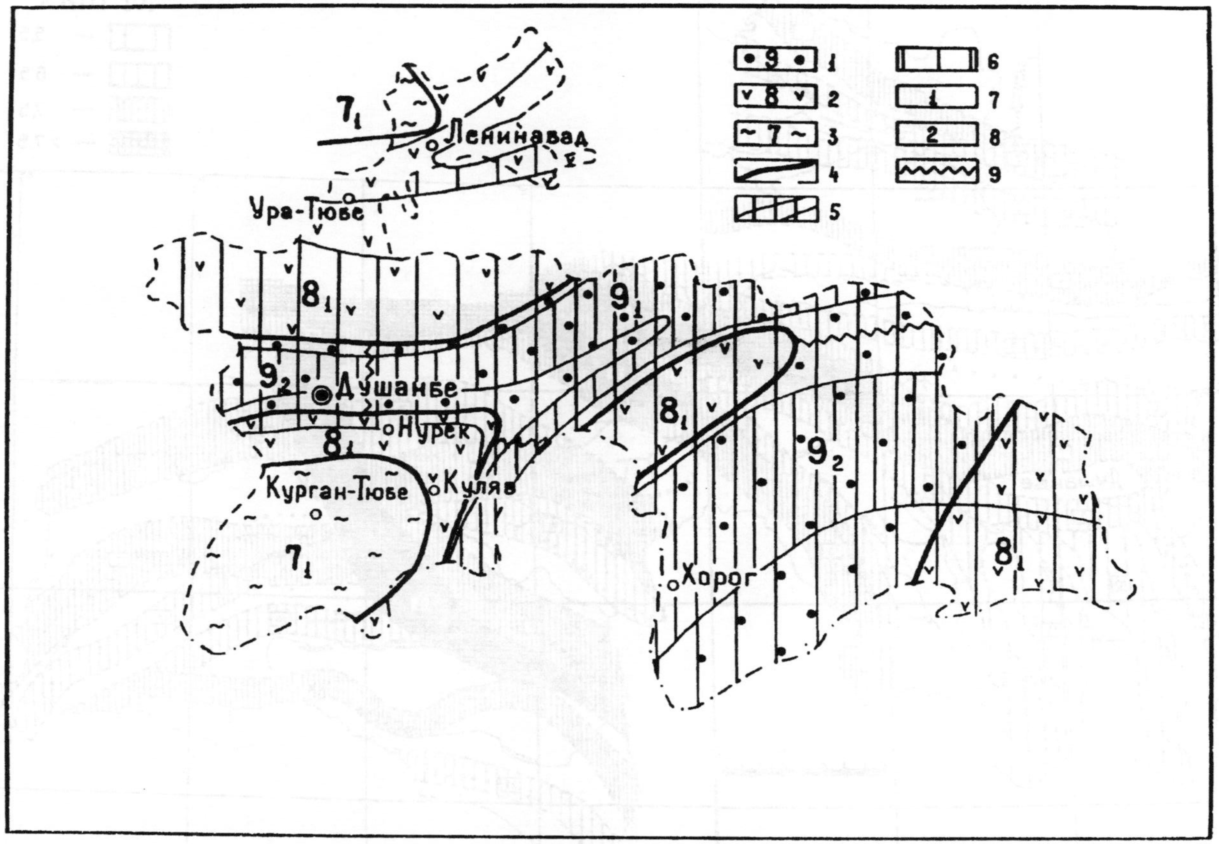

Figure 1. Map of seismic zoning of Tajikistan (1976). 1, 2, and 3 = zones with expected intensity equal to 9, 8, and 7, respectively; 4 = borders of various intensity zones; 5 = zones with M_{max} of 7.5-8.0; 6 = zones with M_{max} of 6.3-7.4; 7 = zones with M_{max} recurrence interval of 30-300 years; 8 = zones with M_{max} recurrence interval of 300-3000 years; 9 = borders of various recurrence interval zones.

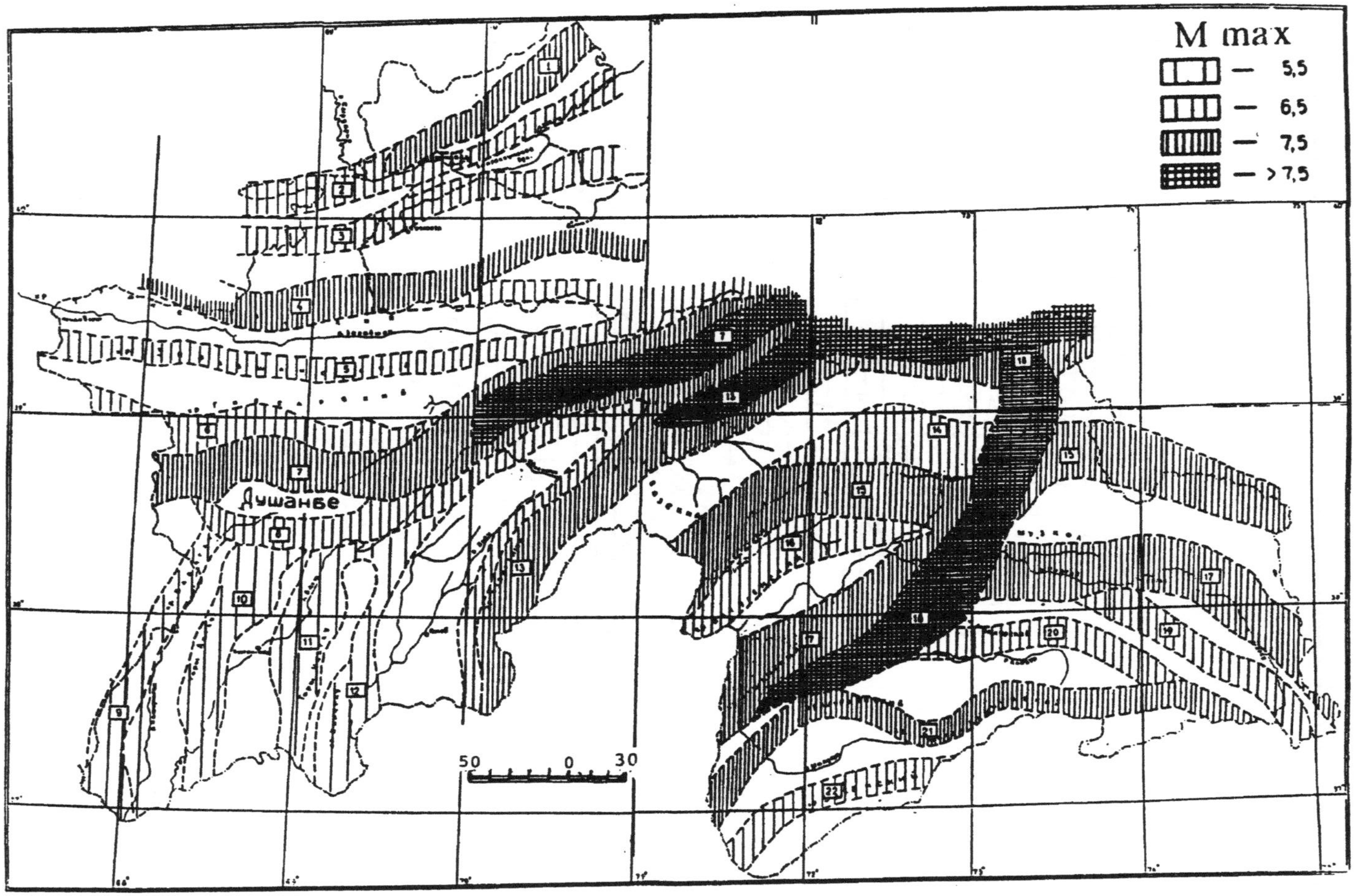

Figure 2. Seismic zones of the Tajikistan territory with four levels of M_{max}. 1 = $M_{max} < 5.6$; 2 = M_{max} of 5.6 to 6.5; 3 = M_{max} of 6.6 to 7.5; 4 = $M_{max} > 7.5$.

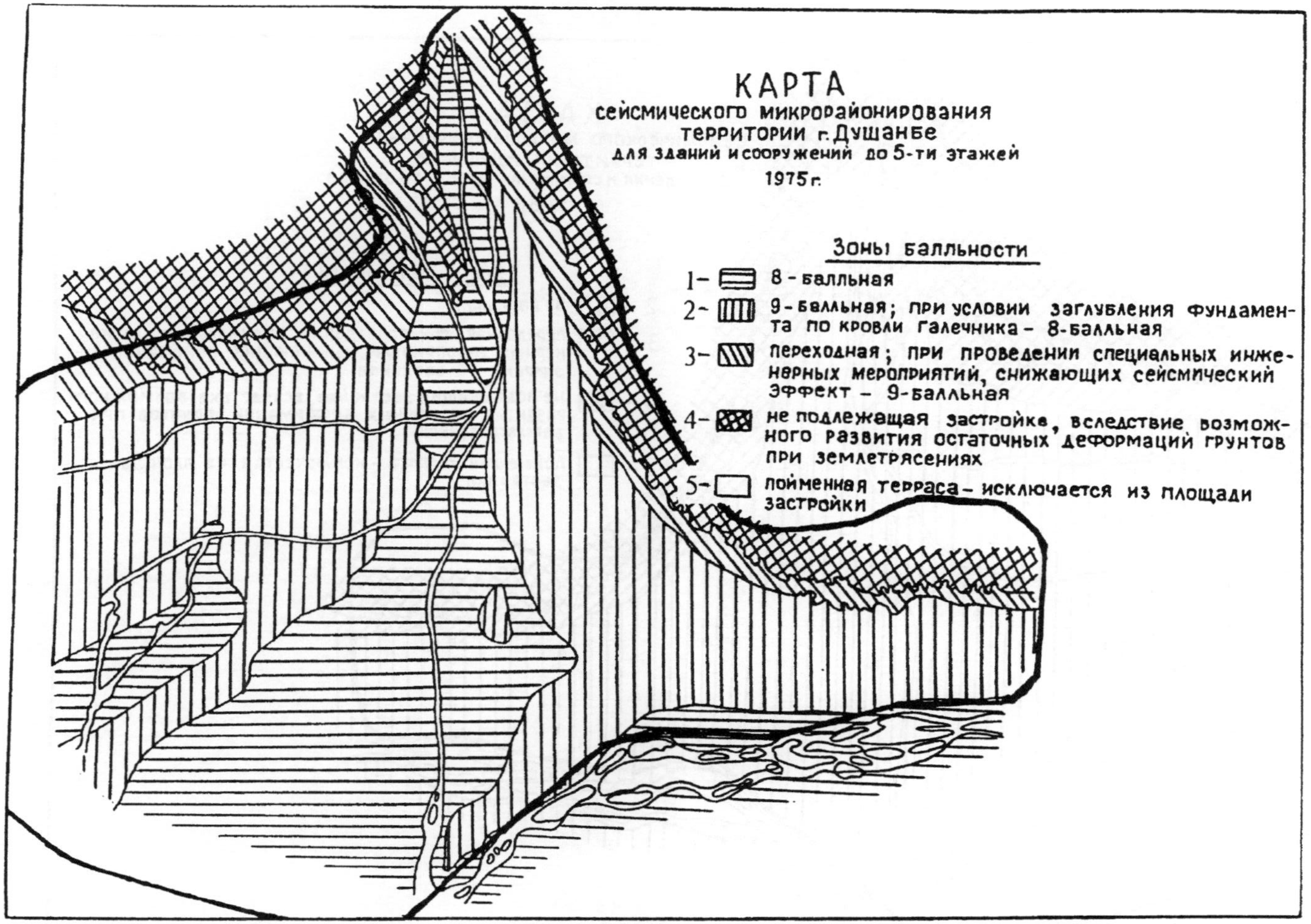

Figure 3a. Map of seismic zoning of the Dushanbe territory for buildings with 5 or less stories.

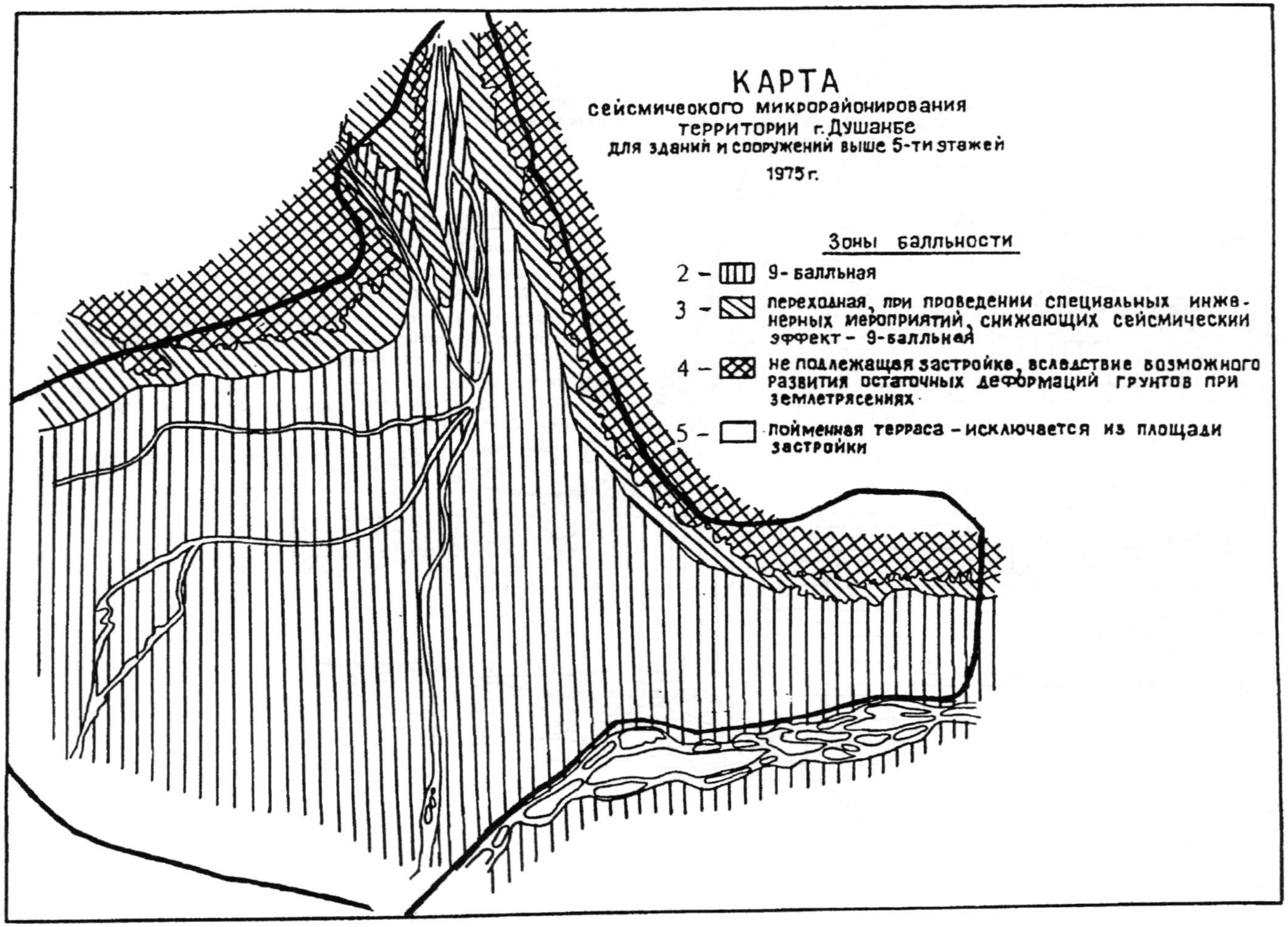

Figure 3b. Map of seismic zoning of the Dushanbe territory for buildings with more than 5 stories.

3. Description of Damaging Earthquakes in Tajikistan

3.1 GENERAL INFORMATION

Macroseismic information about strong earthquakes (with intensity of V or more) exists dating back to the year 1885. Table 1 lists the strong earthquakes that have occurred in the Dushanbe territory.

TABLE 1. Summary of information about strong earthquakes in Dushanbe

Date (mo/da/yr)	Name	Intensity at Epicenter	Intensity in Dushanbe	Coordinates		Magnitude	Depth (km)
				Lat (N)	Long (E)		
10/21/1907	Karatag	IX	VI-VII	38.70	68.10	7.3	24
1/11/1943	Faizabad	VIII-IX	VI	38.62	69.30	6.0	10
7/10/1949	Khait	IX-X	V-VI	39.20	70.80	7.4	16
2/27/1953	Stalinabad	VI-VII	VI-VII	38.80	68.90	4.7	5-10
12/16/1980	Dushanbe	VI-VII	V-VI	38.48	68.75	4.8	---

The number of casualties in Dushanbe due to the earthquakes listed in Table 1 is not known. There were no collapsed buildings in Dushanbe due to these earthquakes. Isoseismal maps of the five earthquakes listed in Table 1 are shown in Figures 4 through 8.

4. Description of Building Construction in Tajikistan

The primary type of building in Dushanbe before 1943 was a mud house constructed from local materials. Later, brick buildings and frame buildings with brick infill walls were constructed. Table 2 list summary data for the various types of residential buildings in Dushanbe. Table 3 lists the number of people living in each type of residential building in Dushanbe.

It should be noted that the most vulnerable buildings, which may experience damage in strong earthquakes, are brick buildings constructed during the period of 1950 to 1970 (series 401 – "Khruschevki"). These buildings comprise approximately 30% of the total residential space. The most reliable buildings are the monolithic reinforced concrete buildings.

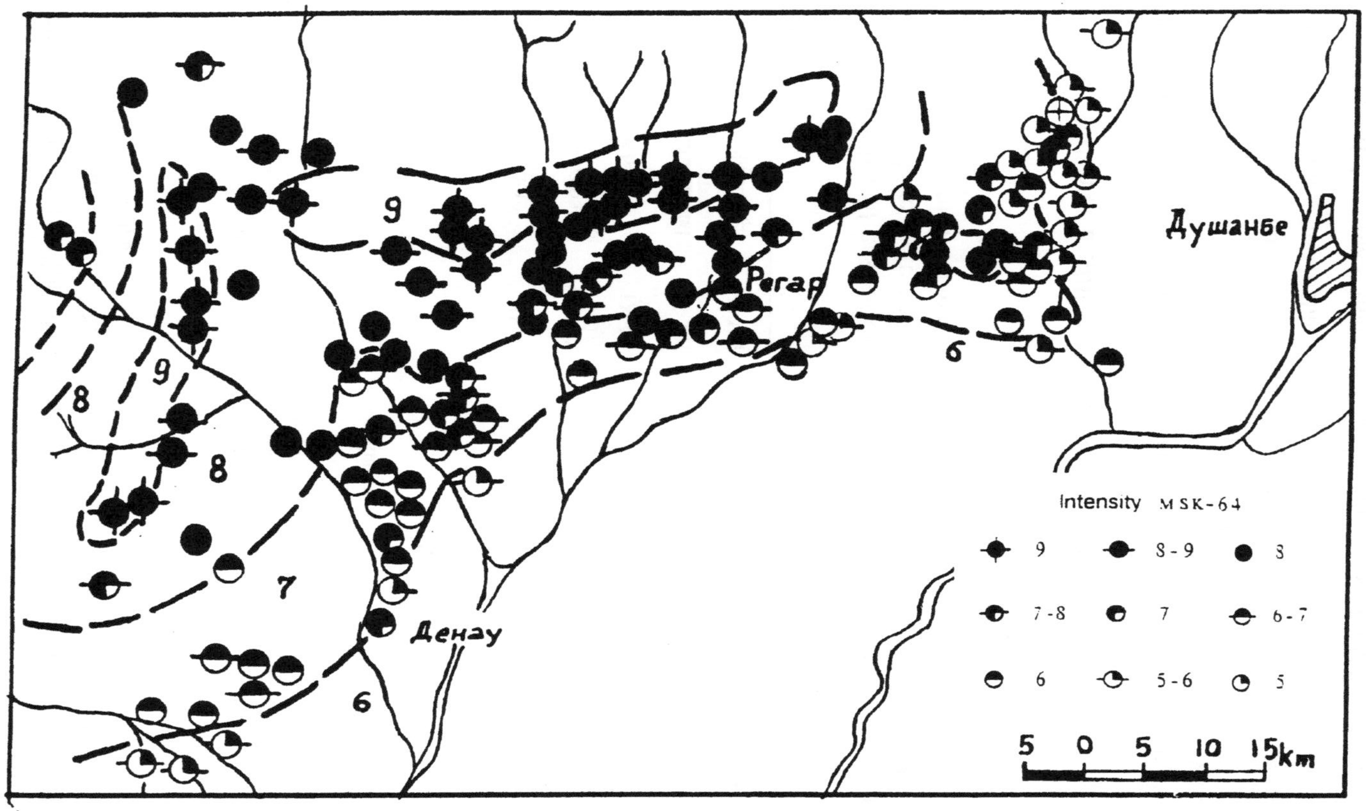

Figure 4. Isoseismal map of the 1907 M 7.3 Karatag earthquake.

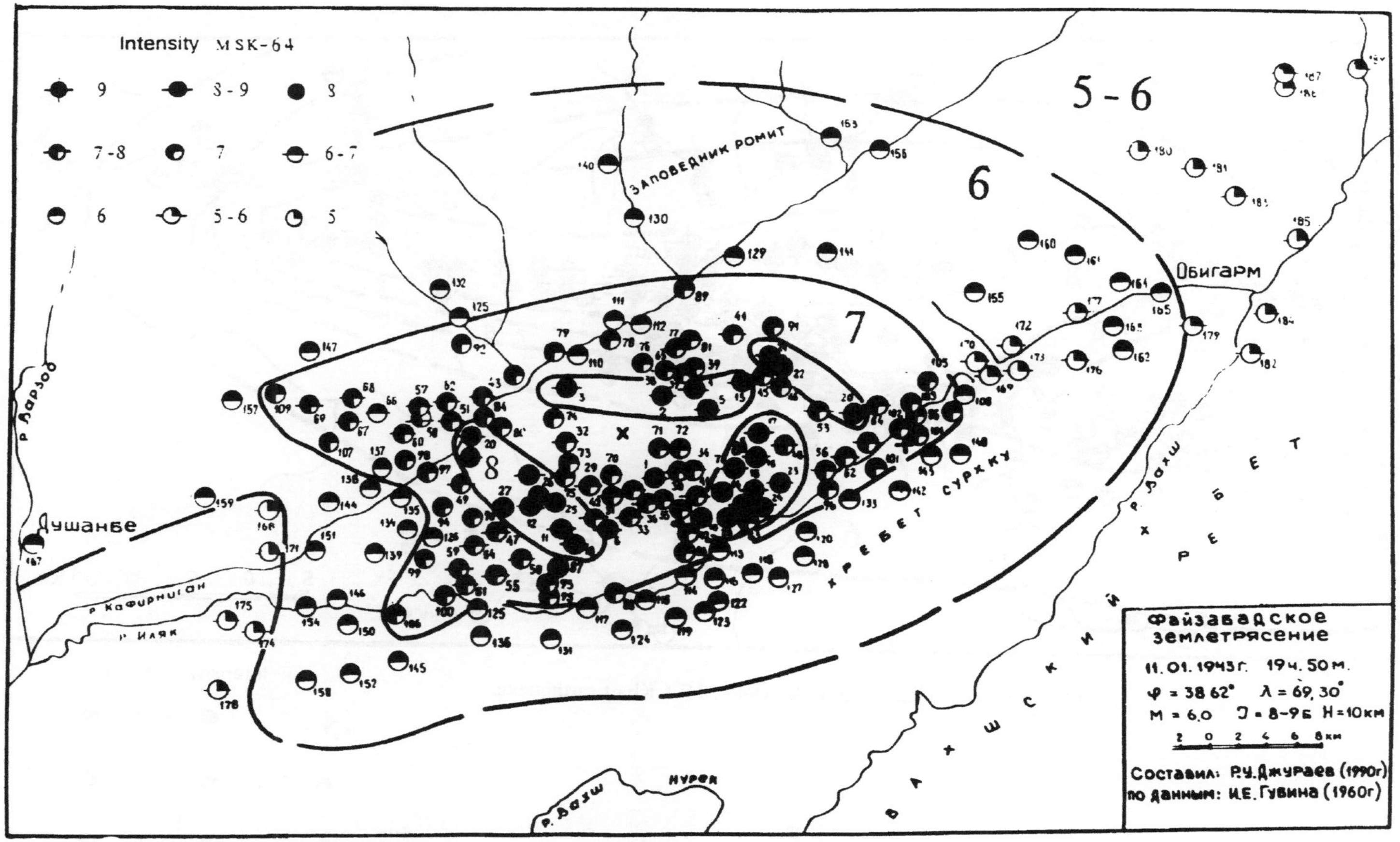

Figure 5. Isoseismal map of the 1943 M 6.0 Faizabad earthquake.

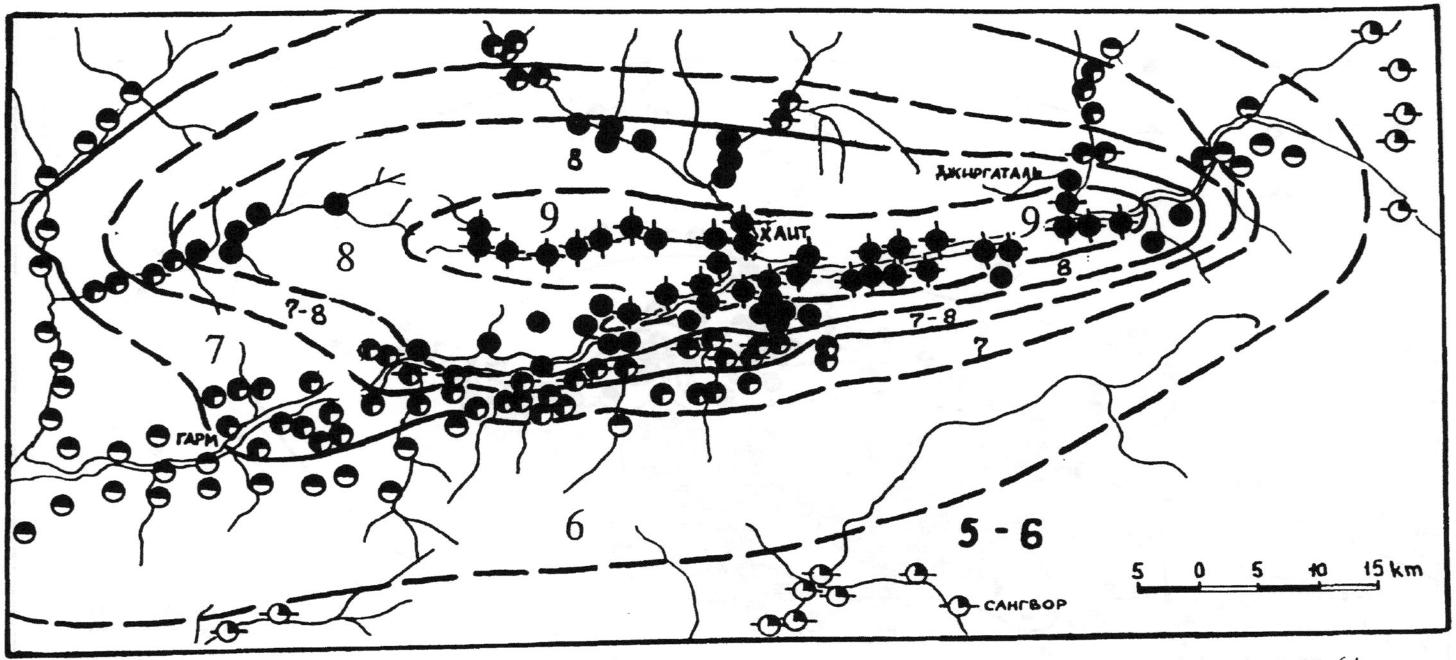

Figure 6. Isoseismal map of the 1949 M 7.4 Khait earthquake.

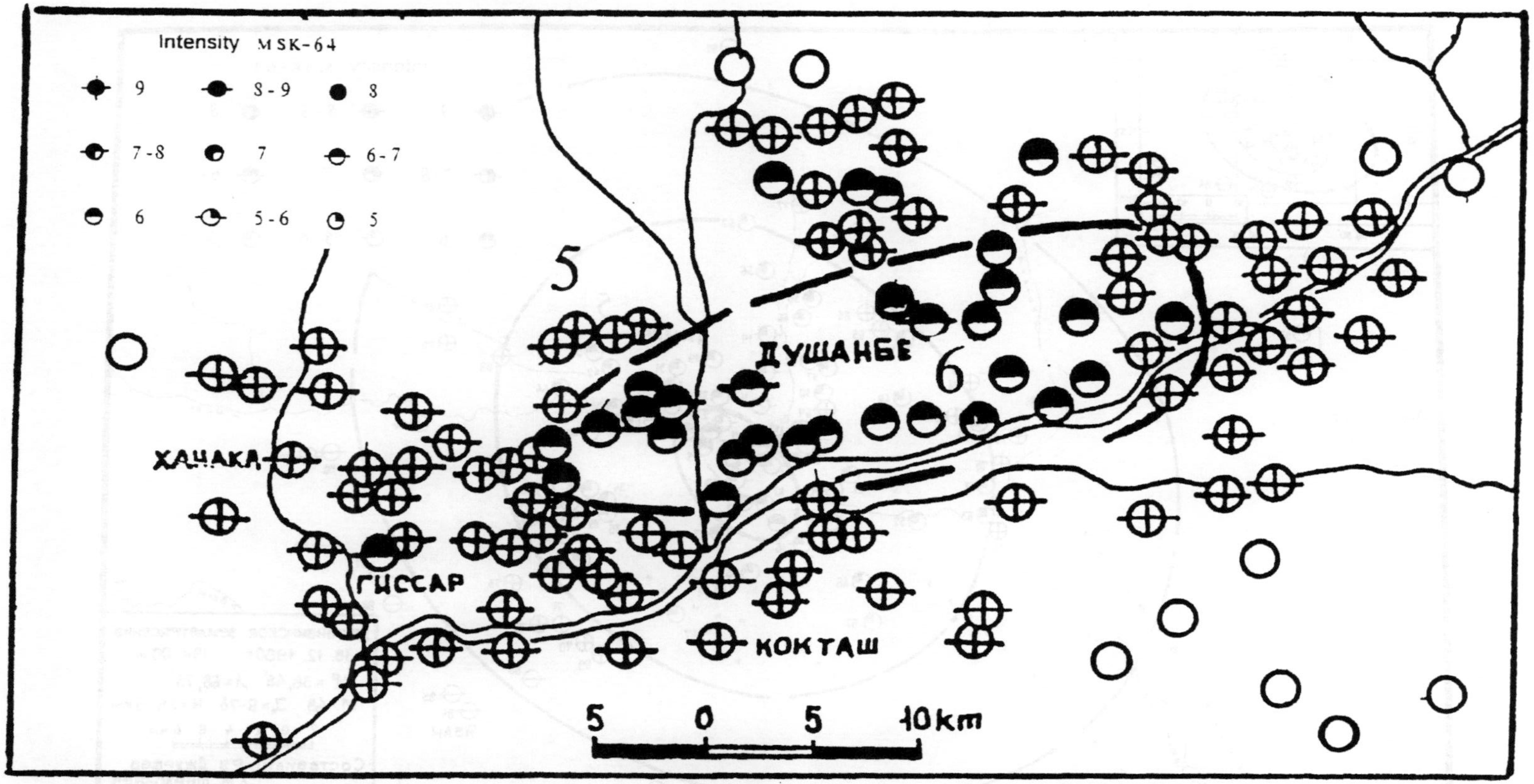

Figure 7. Isoseismal map of the 1952 M 4.7 Stalinabad earthquake.

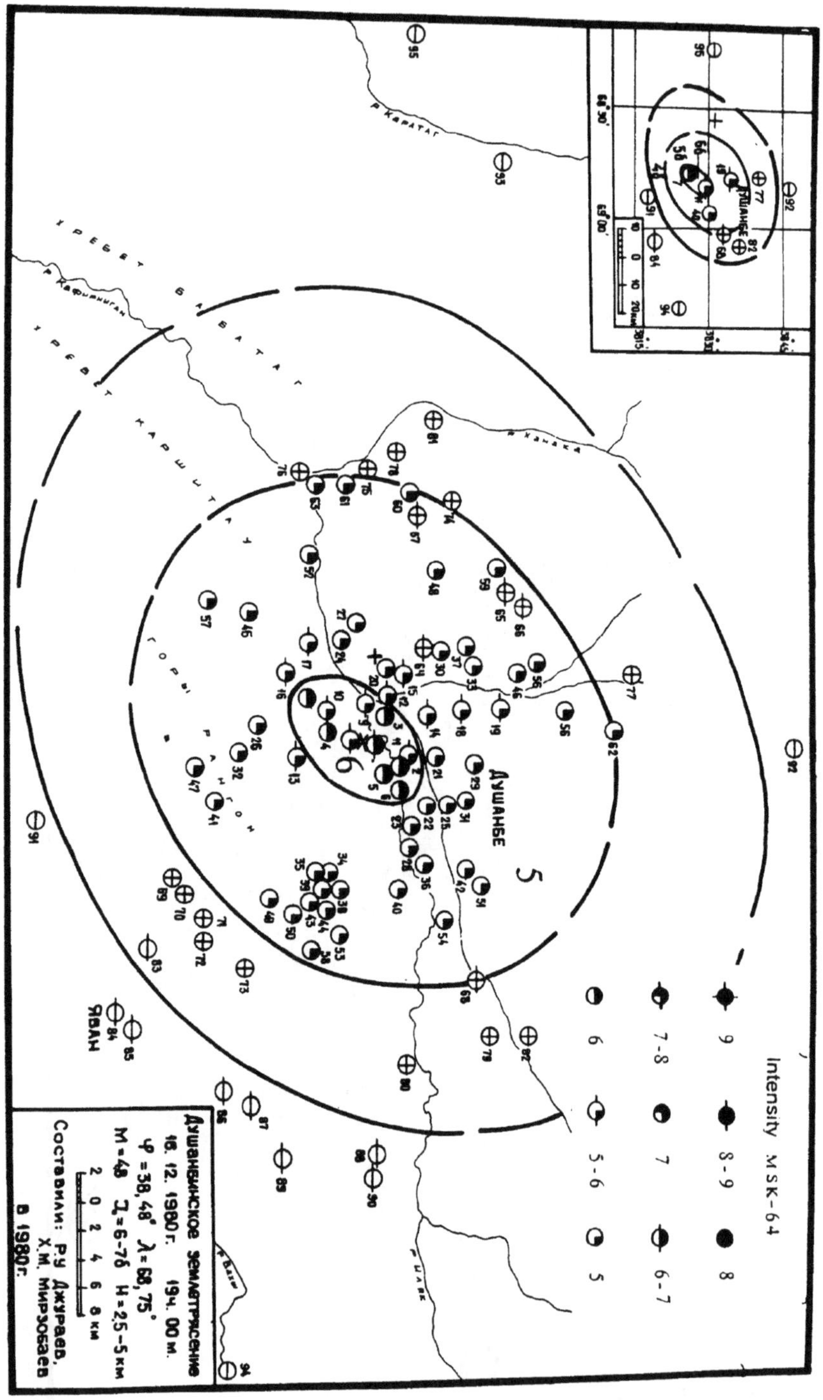

Figure 8. Isoseismal map of the 1980 M 4.8 Dushanbe earthquake.

It should also be noted that the data shown in Table 3 correspond to the time period 1987 to 1989. Now the population of Dushanbe is 1,100,000 people. Due to the events of the civil war in Tajikistan, the calculation of the number of inhabitants in each type of residential building is not possible. However, during the past few years there has been construction of many individual residential buildings without adherence to any seismic design standards. It is estimated that 35% of the population (350,000 people) live in these ramshackle houses. In addition, partial reconstruction of dwellings by adding additional building structure without consideration of support conditions has taken place. These problems may lead to unpredictable consequences in the case of a large earthquake.

An additional problem is that mass construction is taking place in the region of the East Hills in the northwest part of Dushanbe on hillsides with slopes of 15 to 35 degrees and on subsidence loamy loess soils with lenses of subsurface water. In this case, collapse of buildings (cottages of 2 to 3 stories on hillside slopes) is possible both as a result of the development of subsidence deformations and as a result of a strong earthquake with an intensity of VII to VIII. As a result of the Gissar earthquake of 1989, (magnitude 5.5, epicentral intensity of VII to VIII located 25 km from Dushanbe) liquefaction occurred in the loess soils and a very large landslide (3.5 km in length) occurred on a slope of 3 to 5 degrees.

TABLE 2. Summary of building types in Dushanbe (1960-present)

Building Name	Time Period of Construction	Quantity (x1000 m^2)
BRICK with three longitudinal bearing walls		
1. Series TZh I-401; 3-4 story buildings, architect Kh. Yuldashev	1960-1963	---
2. Series I Tzh-401; 4 story buildings	1963-1969	---
3. Series I Tzh-401/69; 4 story buildings (improved)	196-1980	---
4. Series I Tzh-401/69; 4 story buildings with shops on the first floor (first floor is pre-cast monolithic frame)	1970-1990	---
5. Series 155; 4 floor block sections	1979-1986	---
6. Series 155 (improved); 4 story buildings and hostels; 4 story buildings with shops	1986-present	---
7. 4 story buildings of silicate bricks	1960-1970	---

TABLE 2 (cont.). Summary of building types in Dushanbe (1960-present)

Building Name	Time Period of Construction	Quantity (x1000 m^2)
LARGE PANEL		
1. Series ITTZh-464AC; 4 story residential buildings (improved in 1972)	1962-1978	---
2. Series ITTZh-464AC; 5 story residential buildings	1977-1990	---
3. Series 76 (based on series Sh-76/69, 9 story residential buildings and block sections)	1969-present	---
4. Series 76 (based on series Sh-76/68, 9 story buildings and block sections)	1978-present	---
5. Series 105; 5 story block sections	1980-present	6.5
6. Series 105; 9 story block sections	1985-present	---
7. Series 105; 9 story hostel	1988-present	4.0
8. Series 105; 5 story sections	1990-present	---
9. Series 105; 9 story experiment with built in first floor	1983-present	---
BOX-1		
1. Unified frame; without beams but with rigid diaphragm, 9 story residential buildings	1986-present	---
FRAME		
1. Series 118, 4 story residential buildings	1975-present	13.0
2. Individual 9 story residential buildings	1985-present	---
FRAME-PANEL (FRAME IIS-04)		
1. 9 story hostels for students	1983-present	---
2. 9 story hostels for workers	1983-present	---
LIFT-SLAB BUILDINGS (WITH MONOLITH RIGID CORES)		
1. 12 story apartment building with built-in first floor	1980-1988	47.0
2. 12 story apartment building with built-in first floor; walls from hanging panels	1983-present	9.3

TABLE 2 (cont.). Summary of building types in Dushanbe (1960-present)

Building Name	Time Period of Construction	Quantity (x1000 m^2)
MONOLITH REINFORCED CONCRETE (BUILT IN MOVING BLOCK SHIELDS AND SLIDING SHEATHING)		
1. 6 story apartment buildings	1968-1975	8.8
2. 9 story apartment buildings according to Sherozy project	1975-1980	7.0
3. 9 story apartment buildings on Rudaki Avenue	1975-1980	11.2
4. 9 story apartment buildings on Somoni Avenue	1980-1986	11.2
5. 9 story block section building for Kulyab	1982-present	---
6. 12 story apartment buildings; 2 layer outer walls	1983-present	4.0
7. 12 story apartment buildings of Kolkhozproject; inner and outer walls of reinforced concrete	1982-present	---
8. 16 story apartment on Sherozi Avenue; 2 layer outer walls	1983-present	8.0
9. 16 story apartment buildings on Profsoyuz Avenue	1985-present	12.0

TABLE 3. Summary of inhabited buildings in Dushanbe

Type of Building	Total Area (x 1000 m^2)	Number of Inhabitants (thousands)
1. Multistory residential buildings	5656.1	451.87
Monolithic, 16 stories	7.4	0.5
Monolithic, 12 stories	145.0	10.0
Monolithic, 8-9 stories	18.0	1.2
Lift-slab, 12 stories	19.5	1.3
Frame panel, 9 stories	10.6	0.7
Panel, 9 stories	603.0	46.5
Panel, 4-5ories	3875.6	314.47
Brick, 4 stories	770.0	60.0
Hostels, frame-panel	92.0	14.2
Hostels, brick	20.0	3.0

TABLE 3 (cont.). Summary of inhabited buildings in Dushanbe

Type of Building	Total Area (x 1000 m^2)	Number of Inhabitants (thousands)
2. Brick, 2-3 stories	670.0	60.0
3. Individual residential buildings	1130.0	131.28
4. Ramshackle buildings subject to potential damage	7.9	0.69
5. Reinforced residential buildings	36.0	3.16
TOTAL	7500.0	647.0

5. Research Institutes and Organizations in Tajikistan

Table 4 lists the design organizations in Tajikistan that are involved in building design and construction.

TABLE 4. List of projects and design organizations in Tajikistan

Name of Organization	Telephone	Supervising Organization
GTPI Tadzhikprostroi	21-53-71	KOMARKHSTROI of the Government
GTPI TadzhikGIINTIZ	33-95-66	---
GTPI Tadzhikgiproprom	23-08-11	---
SA NIIOSP	33-59-60	---
GPI Tadzhikvodokanalproject	33-52-83	---
PI Dushanbegiprogor	33-92-98	Khukumat Dushanbe
AO Loikha	33-22-72	---
PI Tadzhikkolkhozproject	33-84-24	Tadzhikselstroi
PI Tadzhikgiprotransstroi	21-32-91	Minavtodor
Tadzhikgiprovodkhoz	23-28-21	Tadzhikvodostroi
Kazgiprotorg	21-73-37	---
Tsentrosoyuzprojekt	---	Tadzhukmatlubot
Tadzhikorgtekhstroi	27-19-11	GSK Tadzhikstroi
Tadzhikgorstroi	24-25-45	Umron

TABLE 4 (cont.). List of projects and design organizations in Tajikistan

Name of Organization	Telephone	Supervising Organization
Voyenprojekt	21-73-74	US Minoborony
ORK KPSO Domostroitel	36-05-64	GSK Tadzhikstroi
PSO	21-81-94	TGEStroi
PSB	29-84-95	Aviacpompany Tocikiston
GPI Tadzhikgiprozem	31-16-73	Minselkhoz
Institute of Seismic Resistant Construction and Seismology	25-06-69	Academy of Sciences
Academiya Arkhitektury I Stroitelstva	21-47-17	KOMARKHSTROI
Angishtprojekt	23-25-52	Tadzhikgeologiya
Dushanbearkhproject	21-75-23	Union of Arkh.
TsKTO	36-18-15	Tadzhiklegprom

6. Regulations for Seismic Resistant Design and Construction in Tajikistan

Table 5 gives a summary of the building standards in Tajikistan for seismic resistant design and construction.

TABLE 5. Summary of building standards for seismic resistant construction in Tajikistan

Name of Document	Date Introduced	Document Replaced
PSK-101-51	1951	---
SN-8-57	1/11/57	PSK-101-51; U109-55
SNiP II-A. 12-62	1/3/57	SN-8-57
SNiP II-A. 12-69	1/7/70	SNiP II-A. 12-62
SNiP II-7-81	1/1/82	SNiP II-A. 12-69

7. Seismic Strengthening of Existing Buildings in Tajikistan

Brick buildings are often seismically strengthened by applying shotcrete to the walls. The design is done for a seismic load corresponding to intensity IX. The seismic strengthening of buildings is governed by the branches of the Tajik State Committee of

Construction that deal with existing buildings. Standard documents for strengthening of brick buildings have been tentatively approved by the governing organization.

Currently, the main obstacles to seismic strengthening of buildings are the following:

- Lack of funding for repair and retrofit of residential buildings
- Lack of recommendations and funding for strengthening panel and frame buildings

8. New Approaches to Seismic Design in Tajikistan

At the present time, the organizations that govern seismic resistant design in Tajikistan are not developing new design standards because of the lack of available funding. The existing standards are currently used for building reconstruction and retrofit. There are a large number of individual private construction projects that have taken place without following the current design and construction standards. The primary obstacle to new developments in seismic resistant design is the lack of new standards and guidelines.

9. Optimal Methods for Defining Seismic Loads in Tajikistan

The optimal parameters for seismic load specification are peak accelerations, velocities, and displacements, as well as synthetic accelerograms for given regions. The most important parameter is the peak acceleration. The recurrence interval for defining the expected seismic intensity in Dushanbe should be 100 years.

There are 57 free field strong ground motion recording stations in Tajikistan, and two stations located on dams of the Nurek and Golovnaya Power Stations. In Dushanbe, there are 13 free-field recording stations and 15 stations located in buildings. The maximum recorded acceleration at these stations is 0.6g corresponding to an intensity of VII to VIII.

10. Scientific and Technical Cooperation for Reducing Seismic Risk

Collaboration would be useful in the following fields:

- Engineering seismology
- Assessment of seismic hazard
- Experimental research work on seismic resistance of buildings

  The following items are potential forms of collaboration:

- Creating a unified coordinating organization that includes representatives of all interested countries; this organization would conduct annual meetings or workshops to summarize the results of research work that was carried out during the year

- Sharing of publications, and mutual preparation and publication of reports describing completed research work
- Creating schools or seminars where leading experts would share their training and experience in different branches of seismology and earthquake resistant design and construction

10.1 ISSUES TO BE INCLUDED IN FUTURE WORKSHOPS

The following issues should be on the agenda for discussion at future workshops on seismic risk in Central Asia:

- Defining seismic hazards more accurately (e.g., scales and units of measurement)
- Earthquake resistance of hydro-electric power stations in various regions of Central Asia

10.2 SPECIFIC ITEMS TO FOSTER INTERNATIONAL COLLABORATION

The following items are specific areas where international collaboration would be beneficial in reducing seismic risk in Central Asia:

- Training of specialists in other countries
- Joint work on research projects
- Sharing of information such as materials and data
- Assistance in supplies of equipment and materials needed for research projects

workshop to summarize the results of research work that was carried out during the war.

SEISMIC HAZARD AND BUILDING VULNERABILITY IN TURKMENISTAN

B. KARRYEV
Academy of Sciences, Turkmenistan
B. IL'YASOV
Institute of Seismic Resistant Construction, Turkmenistan
G. GOLINSKY
Institute of Seismology, Turkmenistan

1. Main Data on the Capital City of Turkmenistan

The name of the capital city of Turkmenistan is Ashgabad (sometimes spelled Ashgabat on the figures in this chapter). It is an ancient city, founded in 8000-6000 BC. Its previous names are: Askhabad, Eshkhabad, Ekshabad, and Ishgabad. The city is situated in the central part of the lower mountain plain of Kopetag, on the left bank of a man-made canal at the boundary of the Karakum Desert. The city is located at an elevation of 200-250 meters above sea level, and the geographical coordinates are 37.95° north latitude and 58.30° east longitude.

1.1 SHORT GEOMORPHOLOGICAL DESCRIPTION OF ASHGABAD

The primary form of geomorphological relief corresponds to the main structural elements – wide developments of flat and smooth surfaces and watersheds, and three layers of relief of different ages in the mountain region of Kopetag. The main anticline structures of Kopetag are formed by low Cretaceous deposits made of very firm and strong limestone. That is why the largest positive form of relief corresponds to anticline structures. Synclinal structures consist of soft sand-clay deposits. That is why it often looks as if the river pattern is attached between the lower mountains, although sometimes the rivers cut anticline structures as well.

Ashgabad is situated in the lower mountain plain of Khvalynsky age and envelops the majority of the territory of Central Kopetag, which adjoins the North Ashgabad depression (see Figure 1). South of Ashgabad, in the region of the settlement Yablonovskaya, a wide cone of early or middle quaternary age deposits was developed. North of the settlement, closer to Ashgabad and up the valley, the surface of the cone changes to terraces. Terrace looking surfaces can be seen to the south of the Markou Mountain, which is covered with crushed loamy rock material. North of Yablonovskaya there are Keshenynbairskaya and Pervomaiskaya brakhi anticlines, which began to arise in the second part of the Quaternary period. Within the region of

S.A. King et al. (eds.),
Seismic Hazard and Building Vulnerability in Post-Soviet Central Asian Republics, 127–145.

Predkopetag, an area south of Ashgabad, bending flexure geomorphology that also arose at the end of the Quaternary period is apparent.

1.2 SEISMOTECTONIC DESCRIPTION OF ASHGABAD

The focus zone of Ashgabad earthquakes is tectonically represented by the area of forward fault within the Kalyatinskaya range up to the Gyaurs. This rectilinear alignment in the plan part of the fault differs with the heave displacements at the sides. Intensive abnormal stresses, caused by the transition from one fault to another, arise on the sides. Stretching of the fault in the zones is in the latitudinal direction, perpendicular to the general stress field of the region as shown in Figure 2.

Kopetag meganticlinory arose in the time of the late Mesozoic flexure and exhibits typical epigeosinclinale of the alpine age. A number of abyssal faults divide it into west, central, and east blocks, and differ in morphology and foundation surface and in geologic history. The west block has structural elements of southwest and sublatitude stretching. The central block, within which Ashgabad is situated, differs by the prevailing southeast direction of the structural elements.

The seismotectonic conditions of the Ashgabad region are caused by the location of the city at the foot of the Turkmen-Kharasan Mountains. The city is bounded on the north side by the zone of the Predkopetag fault, and on the south side by the Main Kopetag zone, which divided the region with differing types of flexure during all Yursky time.

1.3 GENERAL SEISMIC ZONING FOR TURKMENISTAN

Figure 3 shows the total map of seismic zoning for the region surrounding Turkmenistan. This map also includes all known isoseismal lines of earthquakes from historical and modern times.

1.4 TOTAL AREA AND POPULATION OF ASHGABAD

The city of Ashgabad stretches 13.5 km from east to west and 12.5 km from north to south. The total territory has an area of about 170 square kilometers. The number of inhabitants was about 545,000 in 1995. The modern architectural appearance of the city was acquired in 1950 after the devastating earthquake of October 5, 1948, in which the majority of structures collapsed.

1.5 MULTISTORY BUILDINGS IN ASHGABAD

More than 40% of the total area of the city is comprised of multistory buildings (3 or more stories). The number of inhabitants in these buildings exceeds 300,000 people.

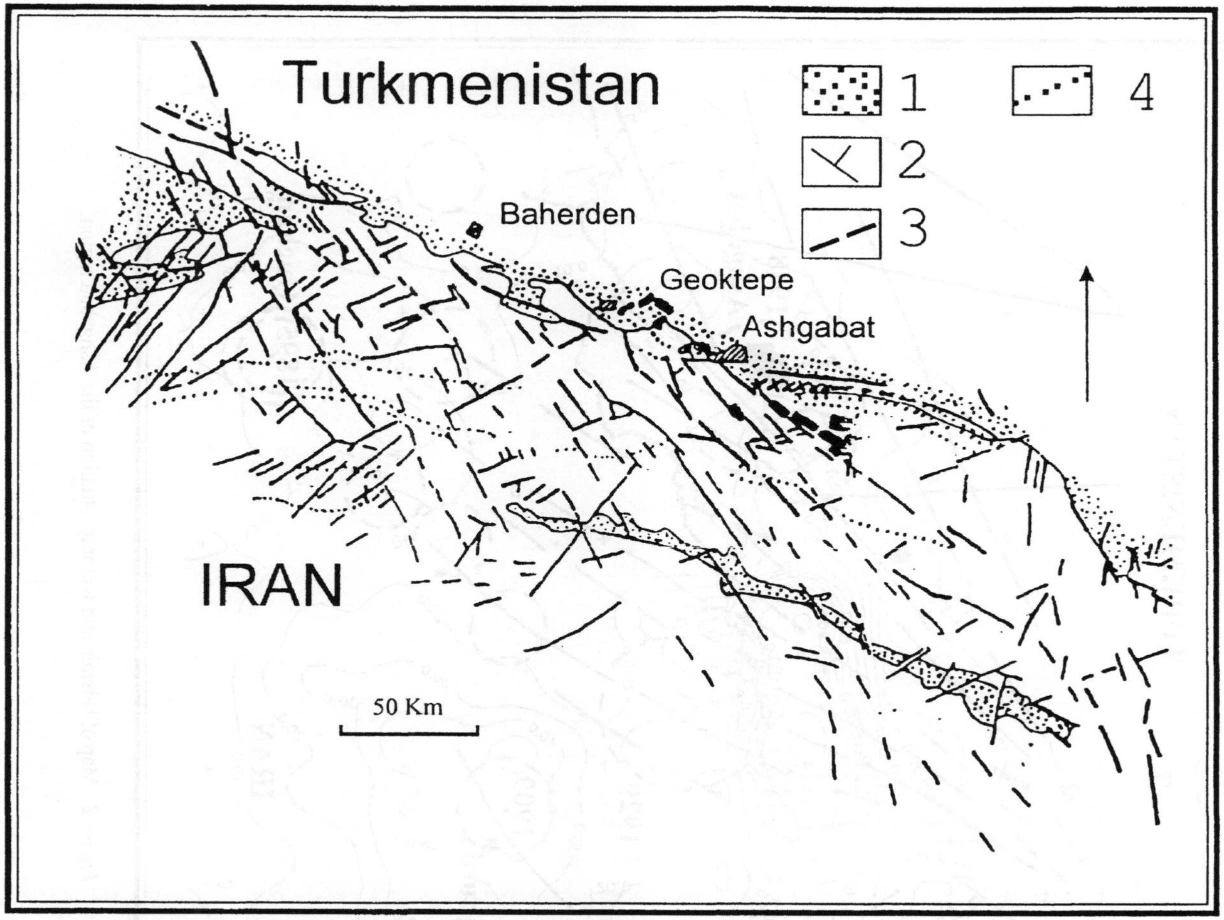

Figure 1. Map of tectonic situation of the Ashgabad region (Rogojin, 1994).

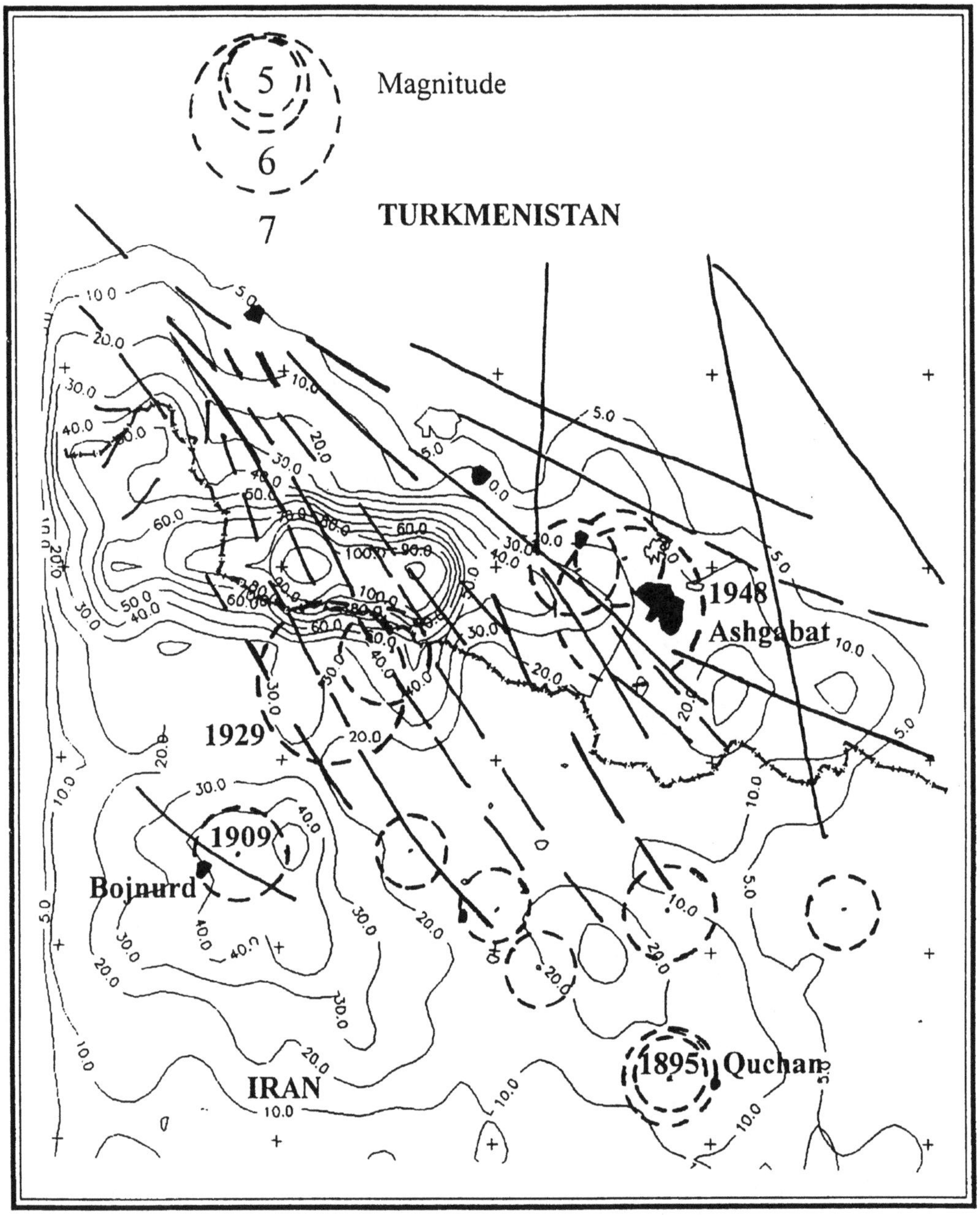

Figure 2. Map of seismic and tectonic situation of the Ashgabad region.

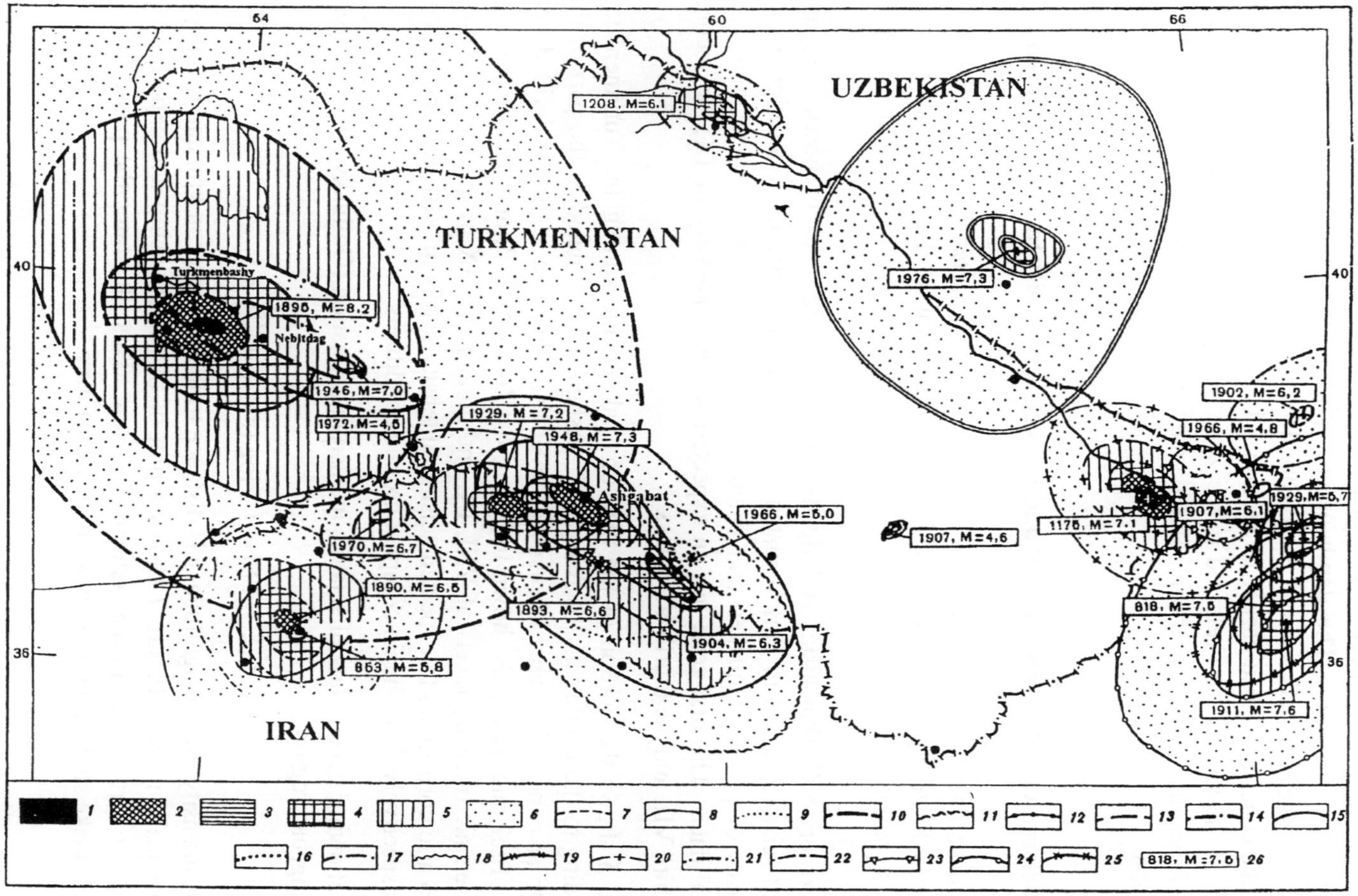

Figure 3. Map of seismic zoning of the Republic of Turkmenistan (Karryev and Galinsky, 1997).

1.6 DESCRIPTION OF SOIL CONDITIONS IN ASHGABAD

The south and west parts of the city have relative good soil conditions in relation to engineering and geologic aspects. The density of soils (crushed stone, sandy loam layers) is about 103 kg/m^3. The depth of groundwater in 1980 was more than 15 meters, but now the level has risen. Less favorable soil conditions for construction are located on the cones described earlier in Section 1.1. These grounds are located in the central part of the city and are referred to the zone with seismic intensity of level IX. The map of seismic zoning has been revised many times. Figure 3 shows the most recent version.

2. Characteristics of Seismic Hazard and Expected Intensity in Turkmenistan

The earthquake hazard is the maximum ground shaking expected in a given location over a specified future period of time. This is very important for Turkmenistan because for the most populated regions, strong earthquake motions are expected in the near future.

The territory of Turkmenistan is divided into five seismic zones as shown in Figure 4. Maps of seismic hazard in Turkmenistan are derived primarily from descriptions of consequences of past earthquakes and calculated seismic effects from tectonic and geophysical data. The first zone has an expected seismic intensity of IX, the second VIII, the third VII, and the fourth VI or less.

Figure 5 shows the portion of the map of total seismic zoning for Turkmenistan corresponding to the capital city of Ashgabad. Ashgabad is situated in the part of the region with an expected seismic intensity of level IX. This area is further divided into six zones of seismic hazard as shown in Figure 6. Zones of 1, 3, and 5 correspond to expected seismic intensity levels of VIII, IX, and greater than IX, respectively. Zones 2, 4, and 6 also correspond to expected seismic intensity levels of VIII, IX, and greater than IX, respectively, but contain complex ground conditions.

The following general law of seismic intensity attenuation (Golinsky, 1977) is used for the Ashgabad region:

$$I_o = 1.5\,M - 3.8 \log (r^2 + h^2)^{1/2} + 3.5 \qquad (1)$$

where I_o is intensity, M is magnitude, r is distance to the epicenter in km, and h is depth of the earthquake in km. Figure 7 shows a plot of this relationship and experimental data from past earthquakes in the Ashgabad region. This illustrates the possibility of strong earthquake shaking in Ashgabad in the near future.

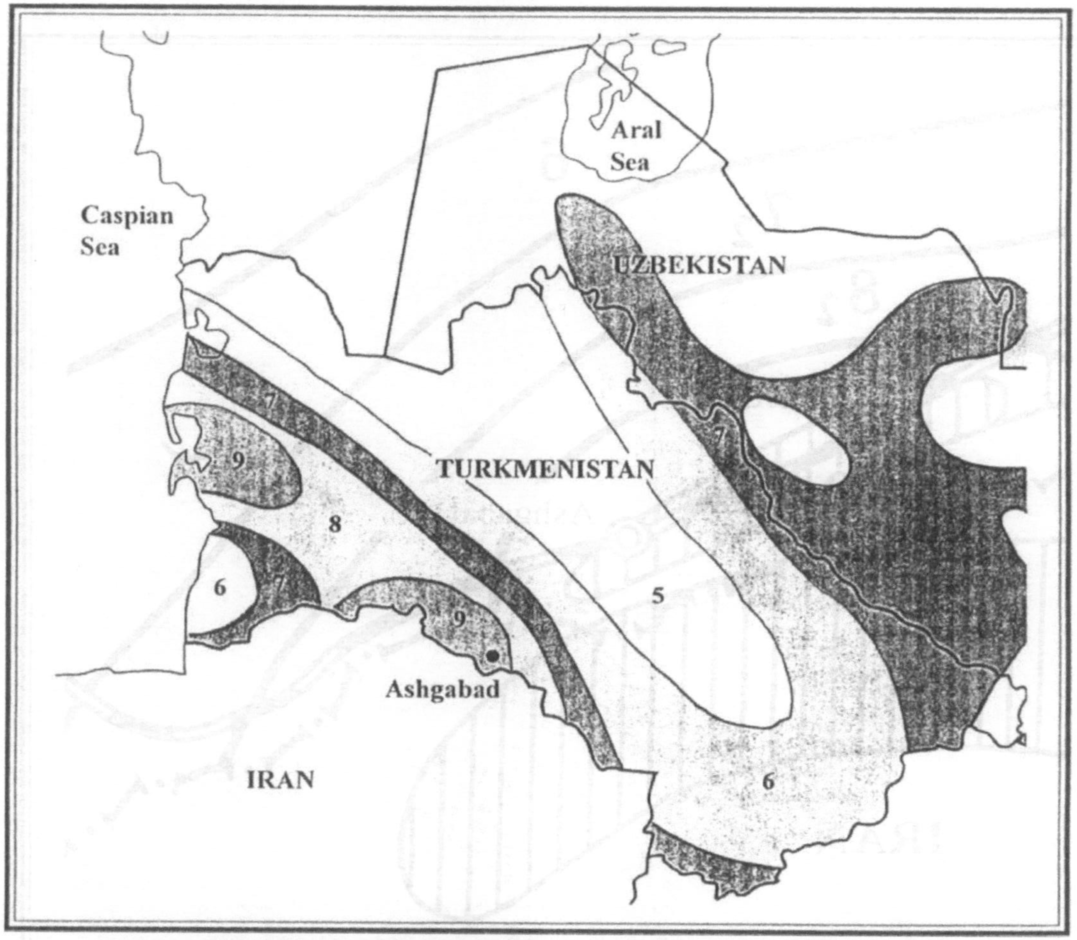

Figure 4. Map of seismic zoning of the Republic of Turkmenistan (Karryev, 1997).

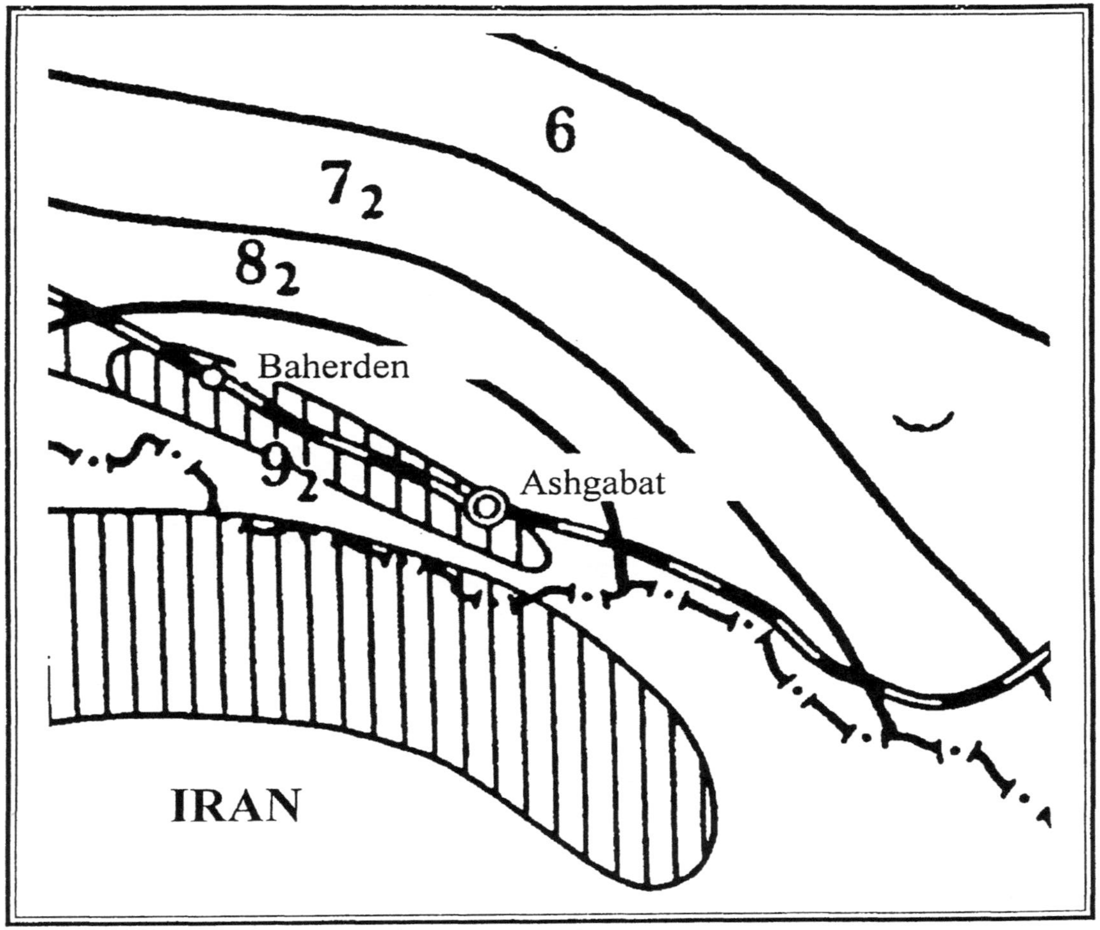

Figure 5. Fragment of map of seismic zoning for Turkmenistan, showing Ashgabad region (Karryev, 1997).

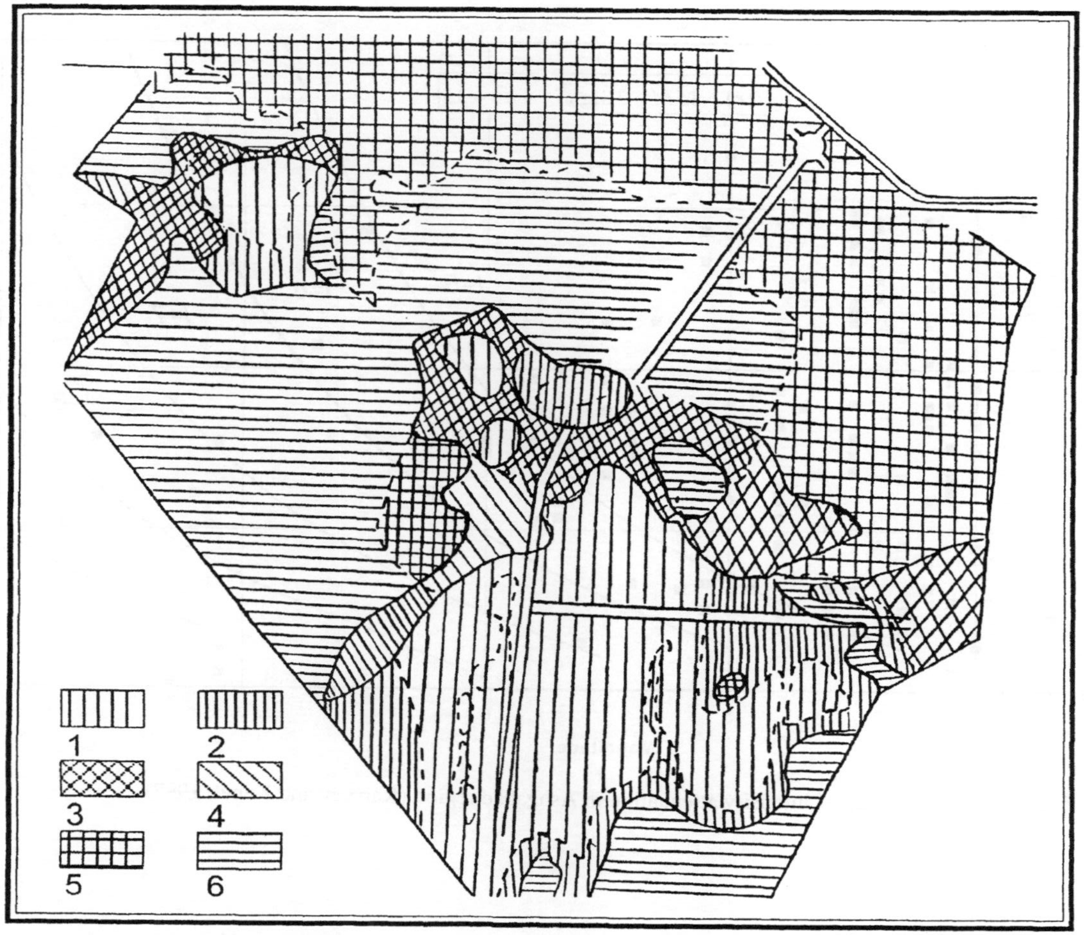

Figure 6. Map of seismic microzoning of Ashgabad region (Vatolin and Esenov, 1984).

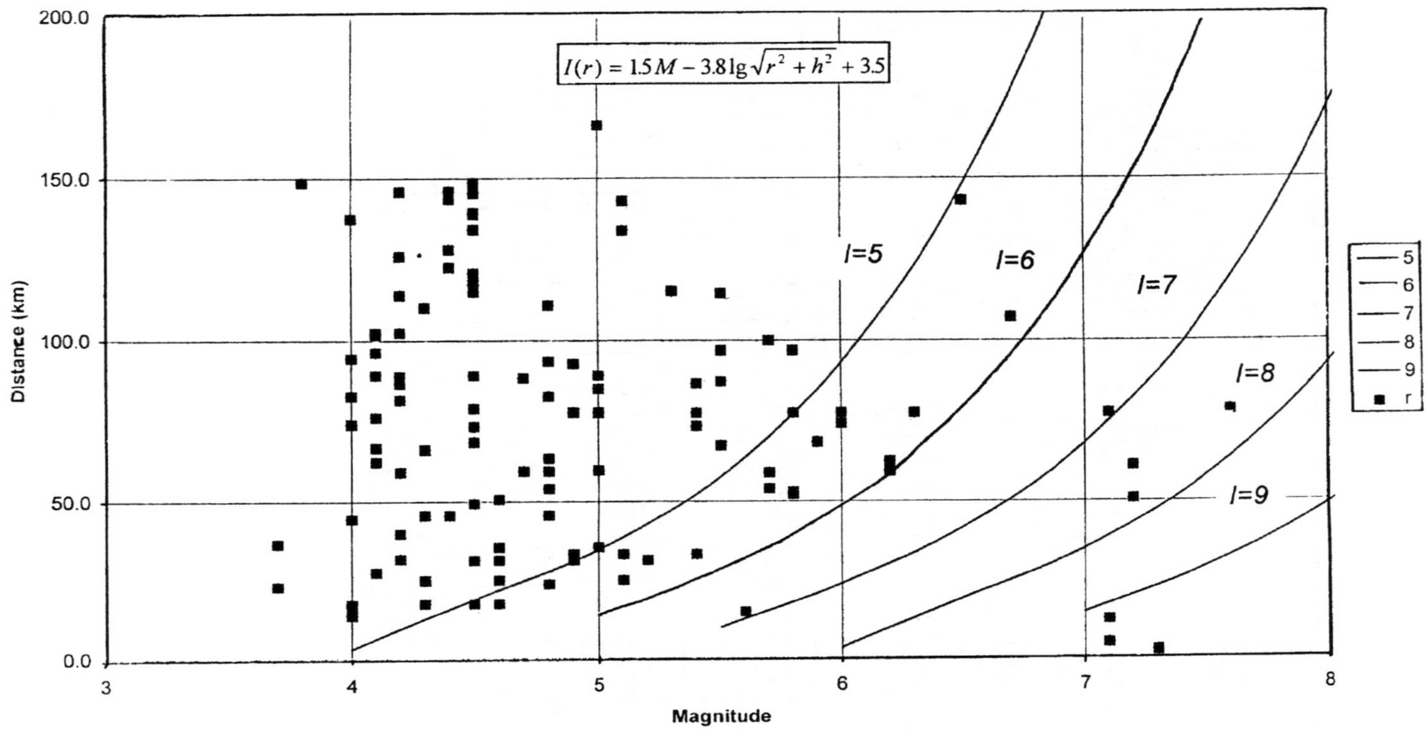

Figure 7. Relationship of intensity attenuation in Ashgabad region (Karryev and Kanel, 1997).

3. Description of Damaging Earthquakes in Turkmenistan

3.1 GENERAL INFORMATION

The most complete catalog of earthquakes along the junction of the Turanian and Iranian plates has been compiled at the Academy of Science of Turkmenistan. The catalog consists of three parts:

- Historical data from 2000 BC to 1900, including about 100 events with magnitudes greater than 5
- Instrumental data from 1900 to 1955 recorded by regional networks outside of Turkmenistan, including nearly 15000 events
- Instrumental data recorded by the Turkmenistan network, including about 54000 events occurring during 1955 to 1995

Table 1 shows a summary of the earthquake catalog for events recorded up to 1992. It should be kept in mind that the low frequency of earthquake events during the earlier centuries is mainly due to incomplete records. The highest number were recorded in the 19th century, when information became more widely available.

TABLE 1. Summary of earthquake data in the seismological database for the Kopetag region

Period (years)	Interval (years)	Intensity V	Intensity VI	Intensity VII	Intensity VIII	TOTAL
Up to our era	2000	0	0	2	0	2
Up to 1000	1000	1	5	5	2	13
1000-1500	500	7	9	10	0	26
1500-1800	300	2	10	2	0	14
1800-1900	100	19	17	4	1	41
1900-1954	54	95	20	3	0	118
1955-1992	38	103	12	5	0	120
TOTAL	3992	227	73	31	3	334

Based on the catalog summarized in Table 1, data for the Ashgabad region was selected and evaluated for completeness with respect to energy class K as shown in Table 2 and seismic intensity as shown in Figure 8.

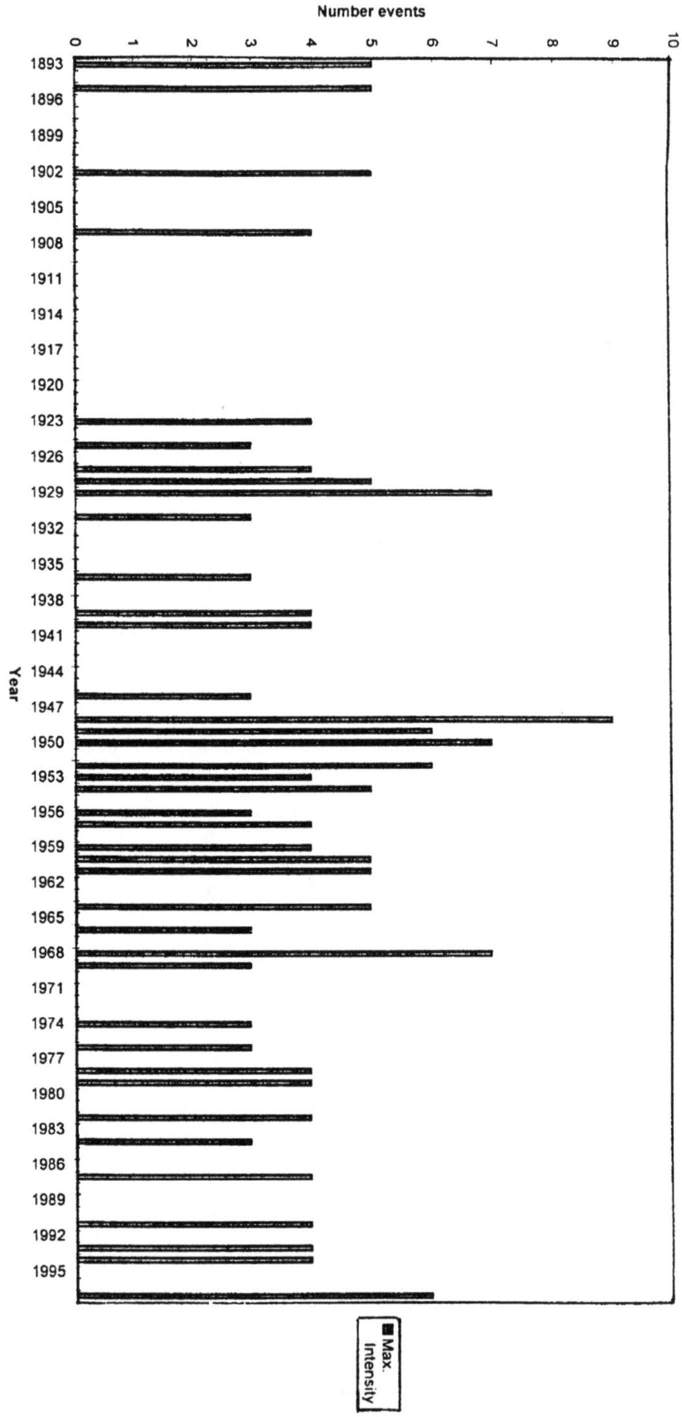

Figure 8. Maximum seismic intensity in the Ashgabad region (1893-1997) (Karryev and Kanel, 1997).

TABLE 2. Summary of earthquake data completeness with respect to energy class K

Energy Class K	Magnitude	No. of Events with Energy Class K	No. of Events with Energy Class Equal to or Greater than K	Data complete since (years)	Period (years)
7 < K < 8	---	5063	7615	1965	30
8 < K < 9	---	3339	3444	1959	35
9 < K < 10	---	1358	1121	1955	40
10 < K < 11	2 < M < 3	420	604	1953	42
11 < K < 12	3 < M < 4	81	185	1950	45
12 < K < 14	4 < M < 5	31	50	1948	47
14 < K < 15	5 < M < 6	8	19	1900	95
15 < K < 17	6 < M < 7	2	2	1800	195

Using a regional relationship of intensity versus magnitude, depth, and distance, as well as information regarding the largest historical earthquakes, four potential local earthquake source zones were identified that pose a severe threat to Ashgabad. The zones are listed in Table 3.

3.2 DESCRIPTION OF 1948 ASHGABAD EARTHQUAKE

Before the 1948 earthquake, the majority of residential buildings were constructed of adobe bricks or they were mud-walled houses. Most of the houses were only one story high and those with more stories had very simple plan configurations. In 1947, there were approximately 8219 residential buildings. Of them, 7973 were one floor adobe brick buildings, 215 were brick buildings up to 3 stories tall, and 31 were wooden frame and wall buildings.

The 1948 Ashgabad earthquake caused the following damage:

- Mud-walled houses and adobe brick buildings: 80% collapsed; 20% had severe damage
- Brick buildings: 25% full collapse; 60% severe damage; 15% partial damage

3.3 REASONS FOR EARTHQUAKE DAMAGE

Possible reasons for damage include the following:

- Unknown seismic hazard in the Ashgabad region and the proximity of the epicenter to the city
- High intensity of short period motions
- Low quality of structural design of buildings
- Low quality of local building materials and lack of seismic resistance

TABLE 3. Information on strong earthquakes (M > 5) in various parts of the Ashgabad area

Zone	Distance (km)	Year	Intensity	Magnitude	Expected Future Intensity
Ashgabad	0	2000 BC	IX	7.1	IX
	10	10	IX	7.1	
	0	1948	IX	7.3	
	0	1968	VII	5.6	
Germab	55	943	VII-VIII	7.6	VIII
	50	1929	VII	7.2	
	55	1948	V	6.0	
	75	1997	VI	7.0	
Bodgnurd	75	1810	V	6.5	VI
	75	1979	VI	6.7	
Quchan	90	1831	V	6.9	VII
		1833	IV	6.2	
		1851	VII	6.9	
		1871	VII	7.2	
		1972	IV	6.3	
		1893	VII	7.1	
		1895	IV	6.0	
		1902	IV	6.2	
		1948	IV	5.9	
		1948	IV	6.0	

4. Description of Building Construction in Turkmenistan

Most of the residential buildings in the inventory of existing buildings constructed after the 1948 earthquake were designed and built based on the seismic requirements of the code at the time of construction. The buildings have seismic resistance measures such as overlapping of monoliths, reinforcement in the brick settings, and seismic joints.

An analysis of building designs shows that their construction meets the requirements of the seismic resistant design code (SNiP II-7-81) for a design load of seismic intensity VIII-IX. Residential buildings of complex structure with brick settings of series IT-395c also satisfy the design requirements. The most seismically resistant buildings are the 4-story large panel buildings of series I-464c, I-V3-500 TSP, and 76. They have a structural system with vertical, longitudinal and cross diaphragms, and walls that form a box structure of high space rigidity. They have proven seismic resistance based on previous strong earthquakes (Gazli and others, 1976). The 9-story large panel buildings of series 1480V have similar qualities, as do the tall buildings of monolithically reinforced concrete, built by sliding and removing the framework.

Brick buildings of series I-195 and I-295c are the buildings which do not have strict construction requirements. In these buildings, the layers of monoliths overlap in widths less than required, and they are constructed of lower quality concrete. There are no inner longitudinal walls along the entire length of the building. Residential buildings of series 263 were designed for seismic loads based on the requirements of static design theory. This does not adequately reflect the actual force distribution between the structural elements, taking into account the elastic properties as they are subjected to earthquake motion.

The actual damage in the city that may occur during a strong earthquake depends both on the structural conditions of buildings and their foundations, and on the local effects of the earthquake caused by the hydro-geological conditions in the region. In the time period 1991-1995, an estimation was made of the possible damage to buildings in Ashgabad. Out of 2463 buildings, it was estimated that 734 (29.5%) can be considered as dangerous buildings (see Table 4).

As time goes by, the seismic situation of the Ashgabad region is becoming more complicated. According to the map of seismic zoning, there are areas with VIII, IX, and greater than IX expected seismic intensity. Currently in the Ashgabad region, there are no design and construction standards in the parts of the region with intensity IX and greater and with complex soil conditions.

5. Research Institutes and Organizations in Turkmenistan

The primary organizations working to develop regional standard documents for regulating seismic design and construction are:

- Academy of Sciences of Turkmenistan – 15, Bitarap Turkmenistan St., Ashgabad, 744000, Turkmenistan

- Research Institute of Seismic Resistant Construction of the Committee for Trsting Construction and Architecture in Turkmenistan – 12, Sad-Keshi, Ashgabad, 744012, Turkmenistan

TABLE 4. Summary of buildings in Ashgabad

Type of Building	Number of Floors	Series	Total Number	Number that are Dangerous
Brick 1	1	Independent Projects	1010	300
Brick 2	2-3	263; 1-195; 1-295	516	323
Complex 1	3	1-395	412	5
Complex 2	4-5	Independent Projects	72	---
Large-Panel	4-5	1-464c; 1-43-500-TSP; 76	405	82
Frame-Panel	4	79	11	---
Frame-Panel	4	Independent Projects	8	---
Monolithic		Independent Projects	29	24
TOTAL			2463	734

Current projects on seismic resistant design and construction include the following:

- Institute Turkmengosproyekt of the Committee for Testing Construction and Architecture in Turkmenistan – 1, Pushkin St., Ashgabad, 744000, Turkmenistan
- Institute Ashgorproyekt, Board on Architecture and City Construction of Khyakimlik – 13, Khivinskaya St., Ashgabad, 744006, Turkmenistan
- Agency Turkmenkommunproyekt, Cabinet of Ministers of Turkmenistan
- Institute Turkmenagroprompoyekt of the Committee for Testing Construction and Architecture in Turkmenistan – 18, Bitarap Turkmenistan St., Ashgabad, 744000, Turkmenistan

Organizations that carry out seismic resistant construction in Ashgabad include the following:

- Committee for Testing Construction and Architecture in Turkmenistan – 56, Navoi St., Ashgabad, 744006, Turkmenistan
- Production Unification Arkachgurulushik – 36, Khudaiberdiyev St., 744000, Turkmenistan

6. Regulations for Seismic Resistant Design and Construction in Turkmenistan

For the time period 1960-1997, the following construction norms and regulations were in force in Ashgabad:

- SNiP II-A, 12-62: "Construction in Seismic Regions" (1962)
- SNiP II-A, 12-69: "Construction in Seismic Regions" (1969)
- SNiP II-7-81: "Construction in Seismic Regions" (1981)

6.1 THEORETICAL BASIS FOR REGULATIONS

The main changes that are reflected in SNiP concern the design formulas for defining seismic loads. Additional coefficients (K1 and K2) were introduced to take into account possible earthquake damage, the structural configuration of buildings, and the construction quality. Also, changes were made in the dynamic coefficient depending on the fundamental period of the building and the seismic properties of the local ground conditions.

6.2 ENFORCEMENT OF DESIGN REGULATIONS

Supervision of design and construction is made by a collective effort of members from design organizations, expertise departments, and quality of construction control of the Ministry of Construction and Architecture of Turkmenistan, Gossandart, and the Institute of Seismic Resistant Construction.

7. Seismic Strengthening of Existing Buildings in Turkmenistan

A large number of buildings in Ashgabad has sustained structural damage caused by differential settlement due to weak soils. The most frequently used method for strengthening existing buildings is the widening of the footings of the building foundation. Other less popular methods include chemically solidifying soils (silicatisation) and cementing saturated sand deposits.

Damaged buildings are restored by traditional methods of strengthening, such as injections into cracks, assembling reinforcement meshed in the layers of plaster or concrete, and adding reinforced concrete or steel rings. Measures for strengthening and reconstruction for individual buildings are developed by the Institute of Seismic Construction using the standard design and construction documents.

Due to the changed seismic situation, in particular the increased seismic hazard caused by rising groundwater levels, more than half of the existing buildings in Ashgabad have inadequate seismic resistance. The high seismicity of the area surrounding the city is taken into account in the design and construction of new buildings, as well as the reconstruction or restoration of old buildings. The seismic design load for strengthening corresponds to a seismic intensity of level IX.

8. New Approaches to Seismic Design in Turkmenistan

The design of new methods is carried out by the Institute Turkmengosproyekt and Ashgorproyekt, Department of Production Unit Arkachgurulushik, Ministry of Construction and Architecture. The regulation documents include "Construction Norms and Regulations on Buildings and Constructions Foundations (SNiP 2.0201-83, - M:1985) and "Design Norms on Construction in Seismic Regions" (SNiP II-7-81, - M:1982; SNiP 2.01.07.85 – Loads, M: 1985).

Design is done for a seismic intensity of level IX. Seismic improvements include increasing the spatial rigidity of buildings and increasing the bearing capacity of the building foundation and the soil on which it sits. The primary obstacle to improved seismic design and construction in Ashgabad is the low quality of local building materials, e.g., the steel for reinforcement and timber for plywood walls.

9. Optimal Methods for Defining Seismic Loads in Turkmenistan

Seismologists have defined 70% of the territory of Ashgabad with a potential seismic intensity of level IX; however, it is also important to define areas with intensity greater than level IX. The type of building construction should be a consideration when defining the seismic intensity level, as should the level of horizontal acceleration that corresponds to the intensity. The spectral method for seismic load definition in building design is relatively simple and appears acceptable for practical use, but it still needs to be tested.

A service for recording earthquake motions on the ground and in buildings has existed at the Institute of Seismic Resistant Construction and the Institute of Seismology at the Academy of Science of Turkmenistan for more than 10 years; however, no recordings of earthquakes have been made in buildings. In 50 locations in the territory of Turkmenistan, 140 records from 79 recordings of the 1983 Kumdag earthquake were made. The maximum recorded acceleration was 300 cm/sec^2 corresponding to a seismic intensity of level VIII. Unfortunately, this network is not currently functioning properly.

10. Scientific and Technical Cooperation for Reducing Seismic Risk

Scientific and technical collaboration among the republics of Central Asia would be very useful. The main topic of collaboration should be seismic resistant construction based on the exchange of experience and research results at conferences and workshops.

10.1 PRIMARY FUNCTION OF A UNIFIED INFORMATION CENTER

It is necessary to create a coordination center so as not to repeat the same research work. The center should be located in Almaty and include a permanent staff of about 5 people. The center should conduct workshops every 2 to 3 years and use the Internet for connecting the center and the republics. In addition, a central database could be established with the use of the World Wide Web.

10.2 ISSUES TO BE INCLUDED IN FUTURE WORKSHOPS

Topics to be addresses at future workshops include the following:

- Seismic resistance of school buildings
- Successful examples of seismic resistant construction
- Development of a local digital seismic network for Turkmenistan
- Creation of an Internet-based data center
- Access to literature and other materials through the World Wide Web
- Possible training of Central Asians in other countries (e.g., USA and Japan)

11. References

1. Golinsky, G.L. (1977) *The Equations of Macroseismic of a Field*, IANT, Ashgabad.
2. Gorshkov, G.P. (1947) *Earthquakes of Turkmenistan*, AS of USSR, Moscow, No 122.
3. Kadirov S.K. (1990) *Ashgabad Earthquake of 1948*, Ilim, Ashgabad.
4. Karryev B.S. (1992) Seismic Phenomena and Structure of Seismic Process in Kopetdag Region, IFZ, Moscow.
5. Karryev, B.S., Ahmedova, S.B., and Golinsky G.L. (1994) Powerful Earthquakes in Turkmenistan *in Seismological Report of Turkmenistan*, Ilim, **2**, pp. 21-48.
6. _____ (1977) The New Catalogue of Strong Earthquakes of the USSR, Moscow.

SEISMIC HAZARD AND BUILDING VULNERABILITY IN UZBEKISTAN

T. RASHIDOV
Institute of Mechanics and Seismic-Resistant Construction, Uzbekistan
L. PLOTNIKOVA
Institute of Geology and Geophysics, Uzbekistan
S. KHAKIMOV
Institute of Typical and Experimental Residential Buildings and Public Buildings, Uzbekistan

1. Main Data on the Capital City of Uzbekistan

The first records about the city Tashkent are found in Chinese literature of the 2nd century BC, in which the Tashkent oasis is called Uni and described as a territory of the state Kangyui. Uni was considered the center of the oasis. In the ancient Persian literature, the name of the Tashkent oasis, Chach, appears in the year 262. Later in the Turkish literature, the city appears as Tash or Chachkent. In the 8th century, the city of Chach was burned and it was revived as the city Shash or Binkent in the 10th century. From the end of the 10th century to the middle of the 12th century, the city was part of the state of Karakhanids and became the capital city named Tashkent, which means "rock city" in the Turkish language. In 1867, Tashkent became the center of a region known as Turkestan Guberniya, and in 1930, it became the capital of Uzbekistan.

The city of Tashkent is situated in the eastern part of Uzbekistan within the Tashkent region, which is a plain crossed by the Syrdarya, Angren, and Chirchik Rivers. In the east and northeast, the city is surrounded by the Naraty range (3000 meters high), spurs of the Chatkal range (4045 meters high), and the Pskem and Ugam ranges (part of the North Tein-Shan). The northeast part of the region, below the mountain ranges, is characterized by sandy clay deposits covered with forests. The plain slopes to the Syrdarya River, and in the east it changes into foothills cut by ravines. The northern part of the plain is called Chirchik-Angren or the Tashkent oasis. The southeast part is known as the Syrdarya oasis, and the southwest part as the Dalverzinskaya steppe. The city is bordered on the south by the Turkestan range.

The geographic coordinates of Tashkent are 41.33° east latitude and 69.25° north longitude.

S.A. King et al. (eds.),
Seismic Hazard and Building Vulnerability in Post-Soviet Central Asian Republics, 147–180.

1.1 SHORT GEOMORPHOLOGICAL DESCRIPTION OF TASHKENT

Tashkent is located within the Pritashkentskaya depression, which is the largest in the mountainous region of Uzbekistan. The depression is located within the Chatcal-Kuramin structural system, between the spurs of the West Tien-Shan mountains in the north, the Karzhantau, Pskem, Urgam, and Chatkal ranges in the northwest, and the Turkestan, Manguzar, and Nuraty ranges in the south. The maximum elevation of the territory is in the northeast part of the Pritashkentsky region at 3277 meters above sea level. The average elevation of the Tashkent depression is about 400 to 450 meters above sea level.

The Pritashkentsky region is referred to as the alluvial terrace portion of the plain, including the valleys of the Chirchic, Akhangaran, and Syrdarya Rivers. The valleys of the Chirchic and Akhangaran Rivers are related to a synclinal fold, which has a width of 2 to 25 km and stretches from northeast to southwest. The sides of the valley gently slope and change into terraces of soil types I, II, and III over flood deposits. The elevation of the surface of the terraces decreases from northeast to southwest and averages about 250 to 300 meters. The width of the terraces is 200 m to 14 km and they rise about 10 to 15 m over each other. Within the valley, the prevailing geology is soil types I and II over flood deposits. The valleys were formed in different epochs of the Quaternary period and are still being formed at the current time.

The relief of the depression is slightly hilly with plains that slope to the southwest in the direction of the Syrdarya River. The relief is crossed by numerous natural Golognaya Steppe and Syrdarya systems of dry valleys and artificial Antropogene canals. In the structure of the Pritashkent engineering-geomorphological zone, primarily proluvial and alluvial Quaternary deposits are present. The alluvial deposits are made up of loess, loess-loamy soils, gravel, boulders, and conglomerates. The total thickness of the Quaternary deposits in the Pritashkent depression is 420 to 500 meters. More ancient bedrock of the Mesozoic and Paleozoic eras can be seen in the outlying parts of the region.

The majority of the city (75% of the area) is located at the surface of the Chirchik-Keles watershed divided by gorges and ravines. The rest of the city (the southern outskirts) occupies the valley of the Chirchik River, with soil types I, II, and III over flood deposit terraces. Within the city, the elevation is less than 140 meters. In addition to the Chirchik River, the city includes a complicated system of artificial and natural canals. The natural relief of the city has been subjected to many changes due to the extensive development of artificial irrigation.

The relief forms of the Tashkent region can be summarized as follows:

- Tashkent proluvial (loess) sloping plain, divided into separate watershed areas by the Chirchic River tributary streams; the average elevation of the plain over the Chirchic River is 35 to 45 meters

- The Chirchic River valley with terraces of three denudacene-accumulative cycles; the second terrace occupies practically all right bank territory in the Tashkent region and stretches about 6 to 7 km in width; it is a flat alluvial plain sloping at 25° and reaching 6 to 8 meters in elevation; the third terrace covers a very limited area within the city

1.2 SEISMOTECTONIC DESCRIPTION OF TASHKENT

According to the scheme of the southwest Tien-Shan tectonic zoning, the Pritashkentsky region is located within the Middle Tien-Shan zone (see Figure 1). The coordinates of the region are 40.00° to 42.30° north latitude and 68.50° to 71.50° east longitude. The tectonic development of the region is characterized by folded crustal structures with rapid motion. The modern structures of the Pritashkentsky region were formed during the long era of alpine tectogenesis. The main structures of the region are the Tashkent-Golognaya Steppe depression and the mountain ranges of the Western Tien-Shan.

The Tashkent-Golognaya Steppe depression has a complicated tectonic structure. It is situated at the boundary of two large structural components and is considered by some researchers (Ryzhkov, Ibragimov, and Yuryev) to be a transition zone between post platform orogeny and the Turnan plate. According to other researchers (Popov, Ryazanov, Rezvoi, and Tal-Virsky), the Pritashkentsky region is crossed by a structural seam, which divides the region into areas with different intensities corresponding to the latest tectonic movements.

The mountain ranges surrounding the Tashkent-Golognaya Steppe depression (Karzhantau, Chatkal, and Kuramy ranges) decrease in the southwest direction and are covered by young structures in some areas (see Figure 2). In the plain part of the depression they are characterized by recent upthrusts and deflections, and are divided by morphocontrolling faults in the northeast direction. The tectonic instability of the region is due to the constant motion of the Tashkent-Golognaya Steppe depression and the Turan plate into orogeny, a region of platform activity, from the alpine time to the present. The tectonic motion rates are 5 to 7 km during Golocene-Antropogene time and 2 to 3 km for upper Pliocene-Antropogene motions.

Strong earthquakes of magnitudes 5.5 to 6.7 have been known to occur in the region since the year 1868. Sixty-seven events of magnitudes 4.5 to 6.7 with intensities VI to VIII (± I) have been observed in the time period of 1868 to the present (see Figure 3). The epicenters of strong earthquakes correspond to outlying regions of large rising features, to the areas with high gradients of recent tectonic motions, and to borders of large structural elements with motions of various directions. Figure 4 shows the map of the latest tectonic motions in Tashkent and its adjoining territories, including the locations of earthquake epicenters in the region.

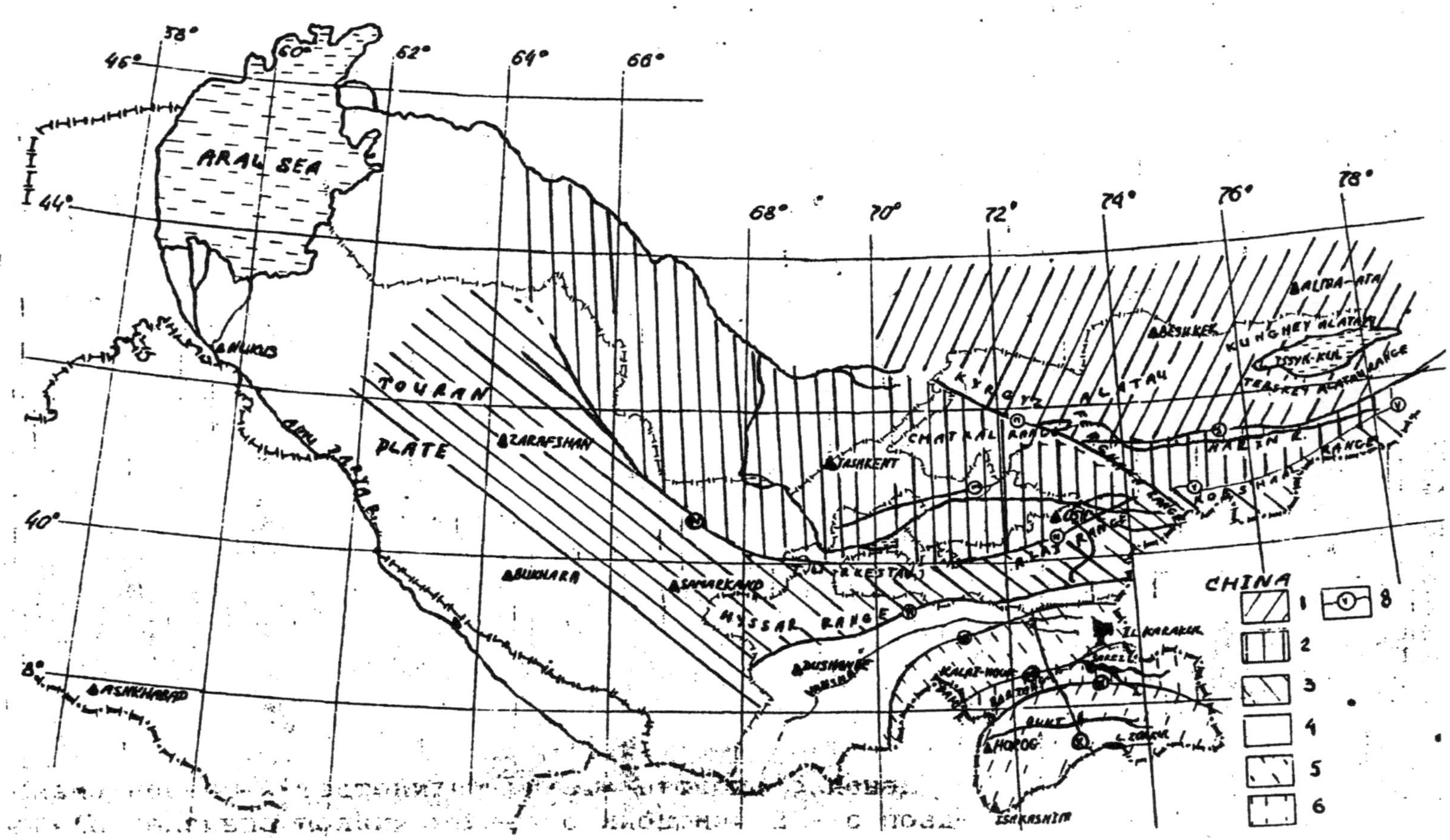

Figure 1. Schematic map of main tectonic units of Central Asia. 1 = North Tien-Shan; 2 = Middle Tien-Shan; 3 = South Tien-Shan; 4 = Alay-Tadjik depression; 5 = Northern Pamir; 6 = Central Pamir; 7 = South Pamir; 8 = Main faults. (I. Chamrabaev, 1996).

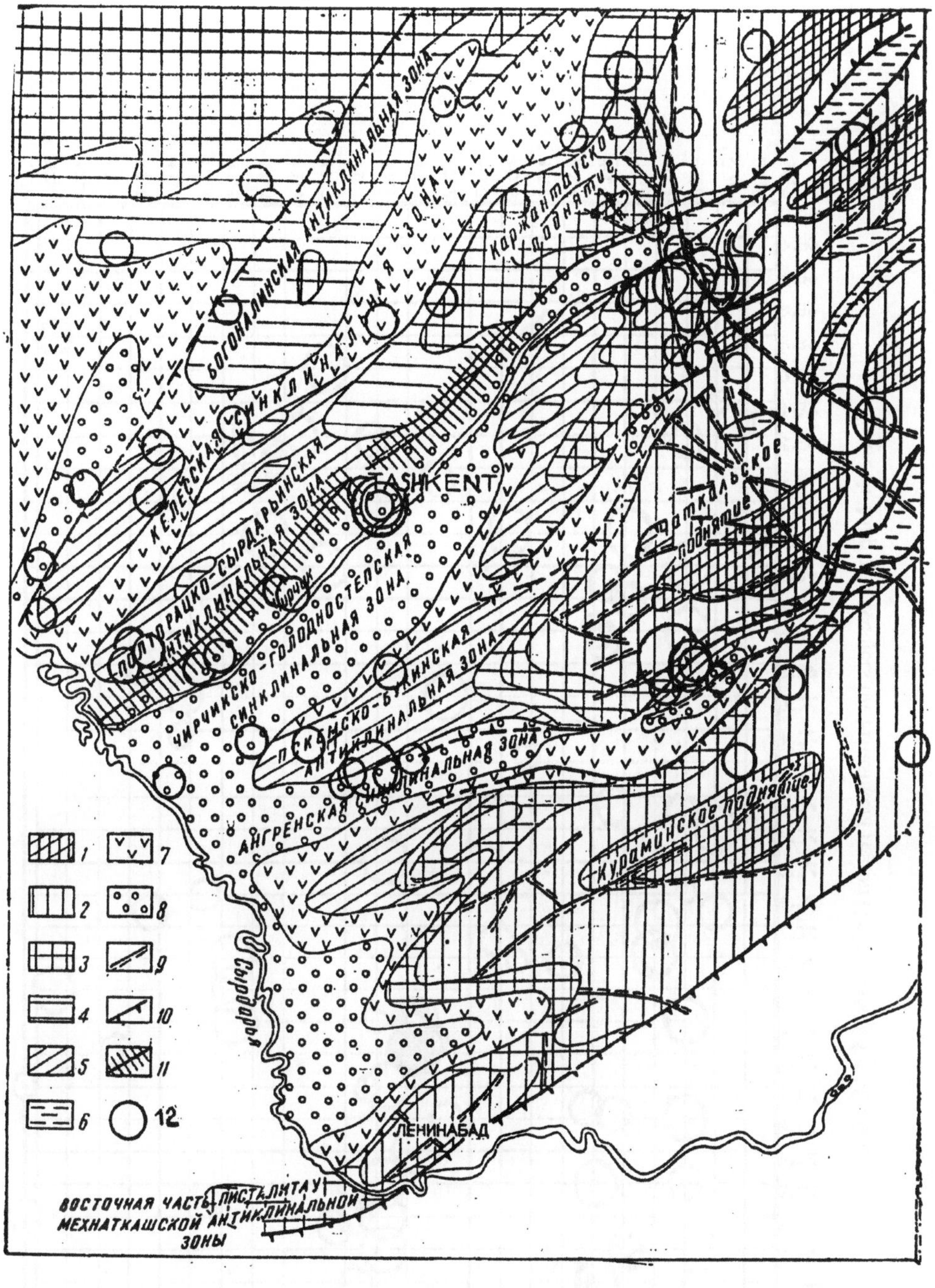

Figure 2. Scheme of recent tectonics of Tashkent region. 1 = Miocene uplift; 2 = Late Pliocene uplift; 3 = Early Pleistocene uplift; 4 = Middle Pleistocene uplift; 5 = Late Pleistocene uplift; 6 = Late Pliocene subsidence; 7 = Middle Pleistocene subsidence and Late Pleistocene uplift; 8 = Recent subsidence.

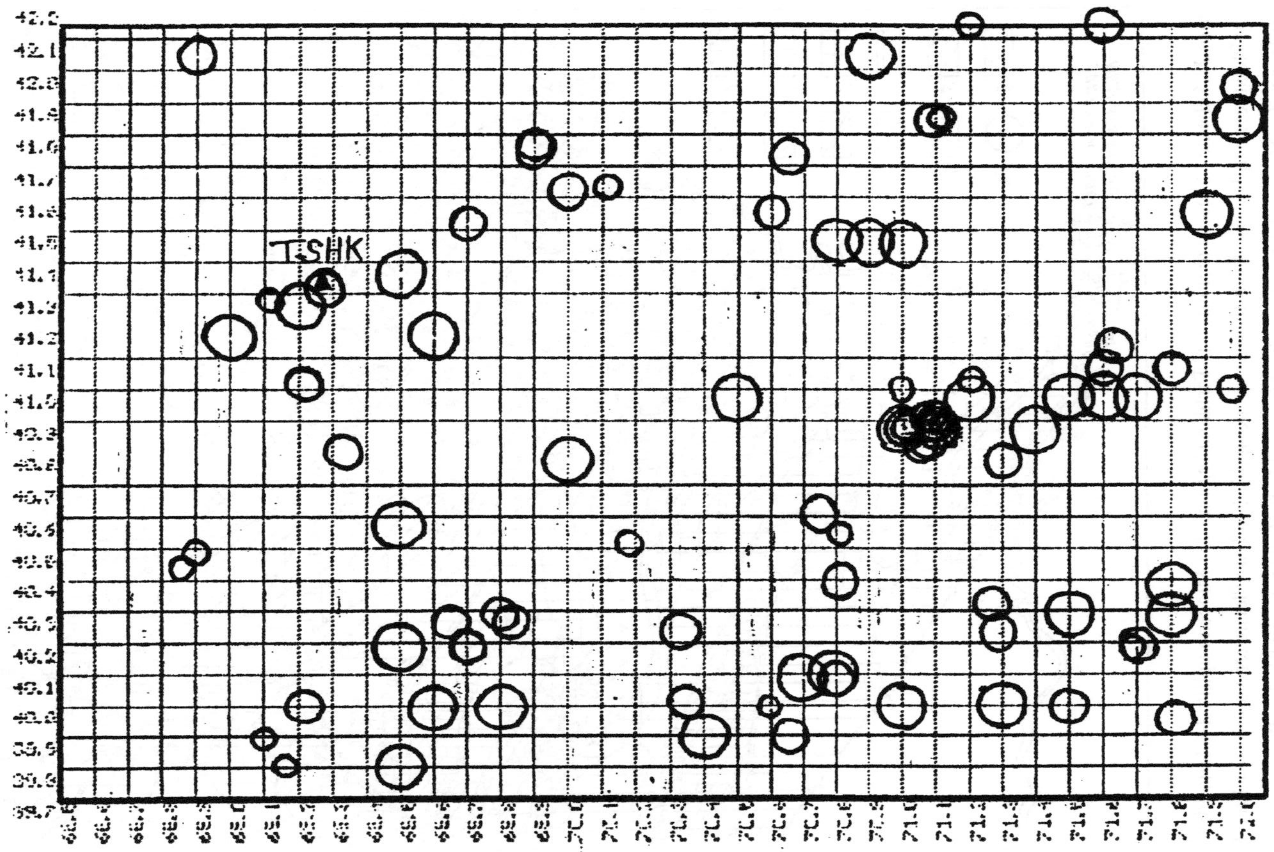

Figure 3. Map of epicenters for Pritashkent region (1868-1996; 4.5 < M < 6.7).

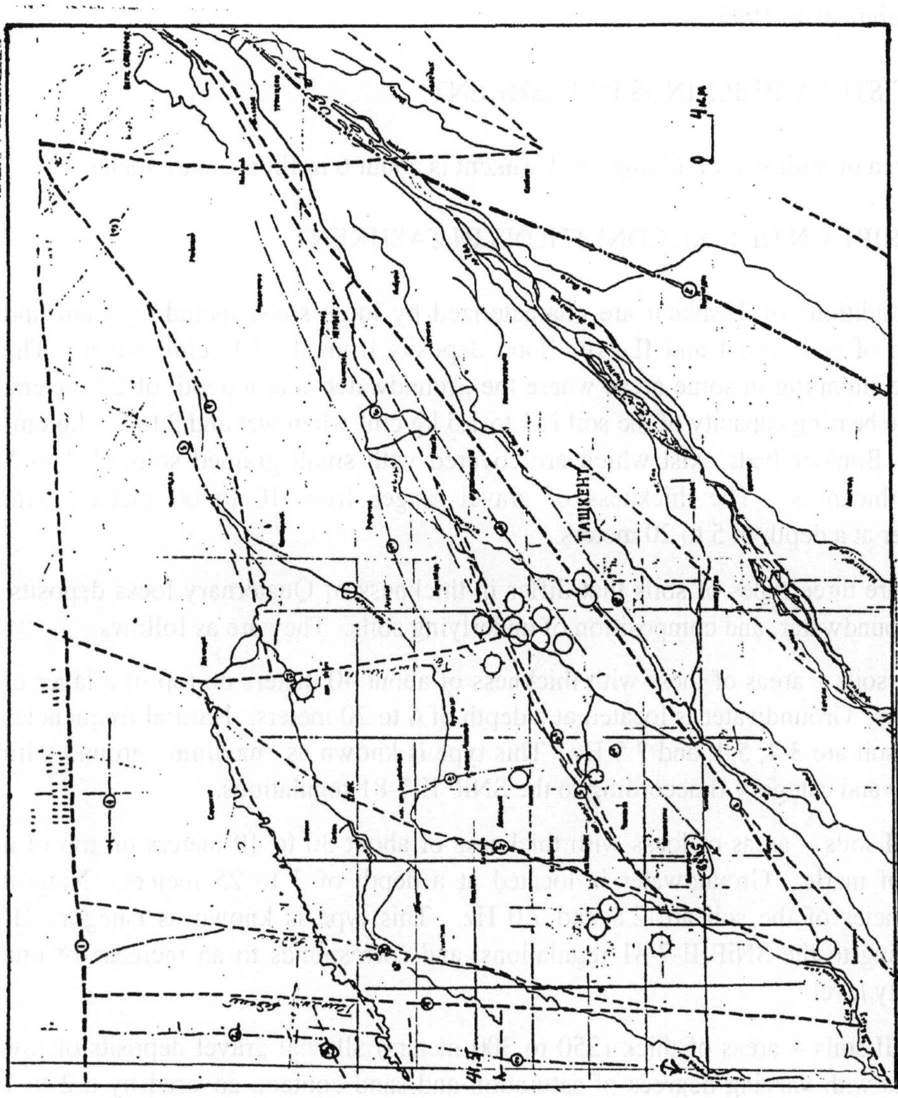

Figure 4. Map of recent tectonics of Tashkent geodynamic polygon (Zakharevich et al., 1982).

1.3 GENERAL SEISMIC ZONING FOR UZBEKISTAN

According to the map of general seismic zoning (GSZ-78) shown in Figure 5, the estimated intensity for the city of Tashkent is VIII.

1.4 TOTAL AREA AND POPULATION OF TASHKENT

The total area of Tashkent is 3030 square km, and the population was more than 2.2 million inhabitants in 1995.

1.5 MULTISTORY BUILDINGS IN TASHKENT

The total area of multistory buildings in Tashkent is about 6 million square meters.

1.6 DESCRIPTION OF SOIL CONDITIONS IN TASHKENT

The soil conditions of Tashkent are characterized by loess soils, including loam and sandy loam of soil type I and II over flood deposits from the Chirchik River. The grounds are slumping in some areas where the groundwater is at a depth of 20 meters. The average bearing capacity of the soil is 1 to 1.5 kg/cm^2 when wet and 2 to 2.5 kg/cm^2 when dry. Boulder beds exist which are covered with small grained soils of 2 to 5 meters in thickness. The thickness of gravel ranges from 10 to 60 meters, with groundwater at a depth of 5 to 20 meters.

There are three types of soils that differ in thickness of Quaternary loess deposits, depth of groundwater, and composition of underlying soils. They are as follows:

- Type I soils – areas of loess with thickness of about 40 meters on top of a layer of boulders. Groundwater is located at a depth of 6 to 20 meters. Natural frequencies of the soil are 3.4, 5.4, and 7.5 Hz. This type is known as "medium" ground with the city and category II according to the SNiP II-7-81 regulations.
- Type II soils – areas of loess with thickness of about 30 to 40 meters on top of a layer of marls. Groundwater is located at a depth of 3 to 25 meters. Natural frequencies of the soil are 2.8 and 7.0 Hz. This type is known as category III according to the SNiP II-7-81 regulations, and corresponds to an increase of one intensity level.
- Type III soils – areas of thick (250 to 300 meters) alluvial gravel deposits of low terraces with varying degrees of saturation and sand content, covered by a 2 to 7 meter thick layer of loess. The natural frequency for this soil type is 2.5 Hz. This type is known as category II according to the SNiP II-7-81 regulations, and corresponds to a decrease of one intensity level for gravel with density more than 1.9 g/cm^3.

The map of seismic microzoning shown in Figure 6 (Kasymov et al., 1984) contains two zones with intensities of VIII and IX. The intensity VIII zone refers to

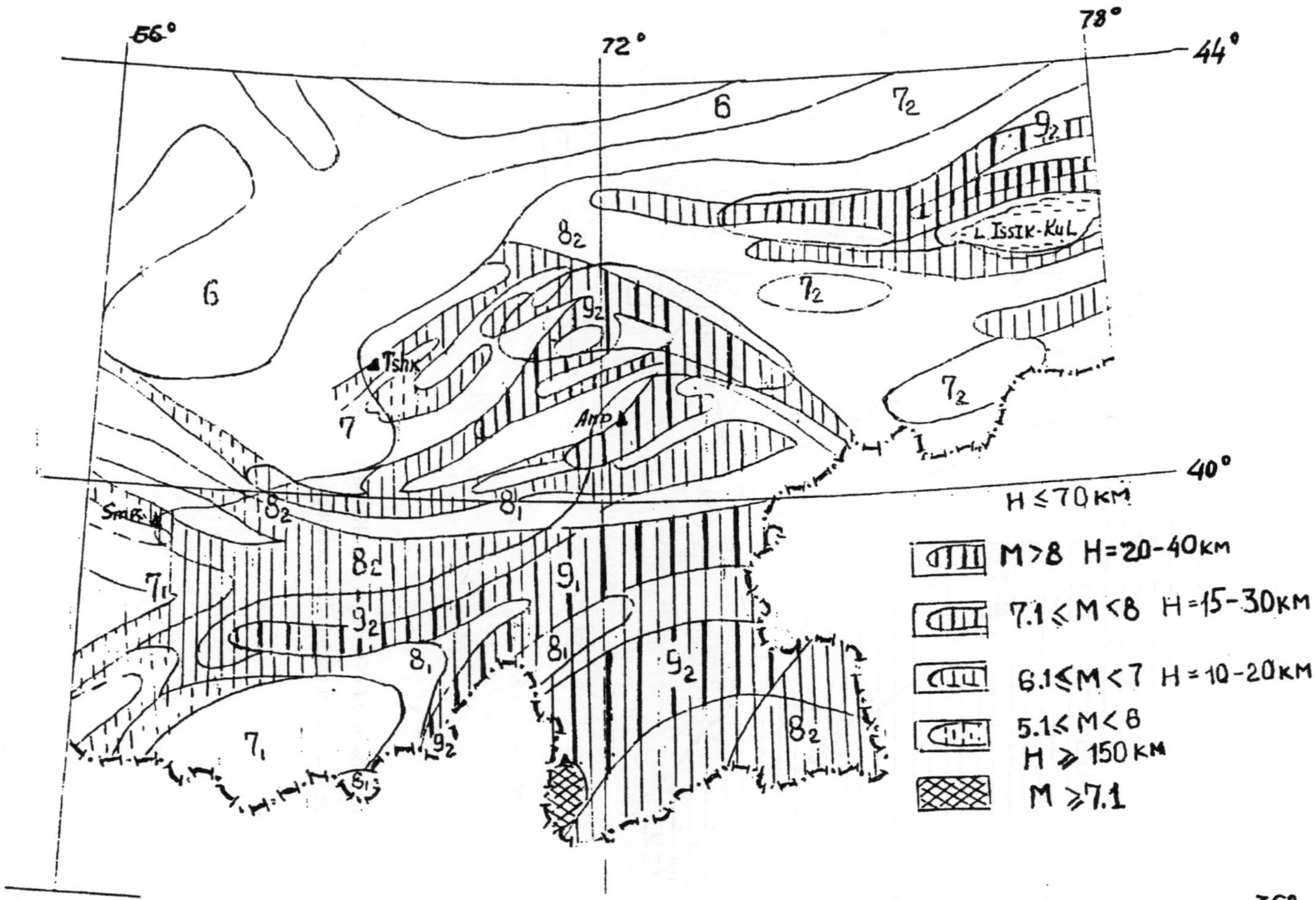

Figure 5. Map of general seismic zoning OSR-78 for Uzbekistan territory.

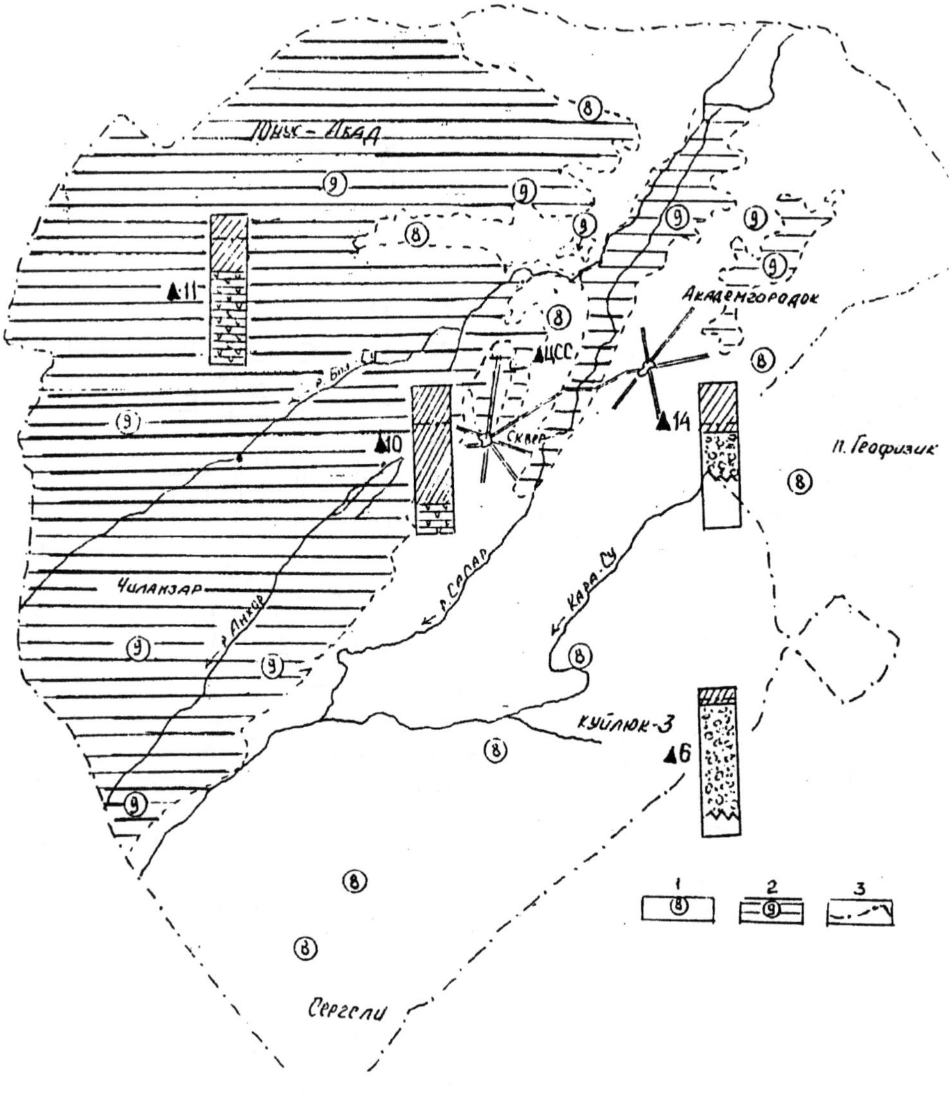

Figure 6. Map of seismic zoning for Tashkent city (Kasymov et al., 1984). 1 = intensity 8; 2 = intensity 9; 3 = level of groundwater; 4 = sites of earthquake recordings.

areas with soils of type I and II, while the intensity IX zone refers to areas with soils of type II.

2. Characteristics of Seismic Hazard and Expected Intensity in Uzbekistan

Figure 7 shows a portion of the seismic zoning map for the Tashkent region. According to the location and age of faults and regions with contrasting motions, the following seismogenic zones connected with morphogenerating structures can be identified (see Figure 8):

- Karzhantau (Pskem-Tashkent) Zone (I) – caused by a series of faults and disjoints of the southwest part of the Tashkent flexural rupture zone. The zone is about 220 km long and 10 to 35 km wide and is characterized by upthrust motion on the order of 1400 to 1500 meters. Satellite images of the northeast part of the Karzhantau fault have shown recent seismic ditches of 2 to 3.5 km in length. About 20% of the total inventory of strong earthquakes in the Pritashkent region can be attributed to this seismotectonic zone. The city of Tashkent is located about 3 to 60 km from this zone.

- Nurekaty Zone (II) – situated to the south of Tashkent at a distance of 20 to 75 km and caused by a fault series in the northeast direction. The zone is more than 100 km in length and 10 to 15 km in width. The mixture and deformations of the terrace deposits of the Angren River show the intensity of tectonic motions in the zone. Earthquakes attributed to this zone include the 1965 M 5.5 Koshtep event, the 1970 M 5.0 Pskent event, and the 1868 M 6.5 Tashkent event. The seismic potential of this zone is estimated at magnitudes less than 6.5.

- Sandalash-Chatkal Zone (III) – located east of zone II at a distance of 160 to 220 km from Tashkent. There are a number of paleodislocations and residual deformations within this zone. The 1946 M 7.5 Chatkal earthquake is attributed to this zone. The seismic potential of this zone is estimated at magnitudes less than 7.5 and intensity levels of IX or less.

- North Fergana Zone (IV) – caused by the North Fergana fault and its junctions with the Kumbel, Kenkol, and Bashtaven faults. Tashkent is located about 130 to 180 km from this zone. Earthquakes attributed to this zone include the 1888 M 6.3 Kostakoz event, the 1894 M 5.6 Chust event, and the 1985 M 6.1 Karakum event.

- Bogonali and Mansuraty Zones (V) – located in the north part of the Pritashkent region and caused by the Bogonali and Mansuraty faults. The amplitude of Quaternary movement on these faults is 500 meters of less. Tashkent is located about 120 to 150 km from this zone. Only weak earthquakes of magnitudes less than 2.5 are expected in this zone.

- Kumbel-Ugam Zone (VI) – caused by deep regional faults of northwest orientation. This it typically a low seismicity zone, but a number of paleoseismic dislocations show potential for magnitude 6.1 to 6.7 events with an intensity of IX. Tashkent is located about 60 to 110 km from this zone.

In the Tashkent region, seismic intensity levels of VII or more may be caused by earthquakes from zones I, II, or III. The map of main faults in the region is shown in Figure 9 (based on Dzhamalov), and the seismic parameters of the most active of the six zones are shown in Table 1.

TABLE 1. Seismic parameters of earthquake zones in Tashkent region

Zone Name	Maximum Expected Magnitude	Maximum Expected Intensity	Source Mechanisms
Karjantau	6.8	IX	Thrust; compression across structures; dilatation along structures
Nurekata	6.4	VIII ½	Thrust-slip; compression across structures; dilatation along structures
Sandalash-Chatkal	7.5	IX	Thrust
Kumbel-Ugam	5.8	VIII	Thrust; compression across structures; dilatation along structures

Table 2 shows statistics on the recurrence of earthquake events in the Pritashkent region (area equal to 55,000 km^2) according to observed data from 1868 to 1995. Table 3 shows the return periods for various intensity levels in Tashkent, also based on observed data for the 1868 to 1995 time period. According to the general map of seismic zoning (GSZ-78), the maximum expected seismic intensity in Tashkent is VIII with a return period of 1000 years (a probability of 0.05 in the next 50 years).

TABLE 2. Earthquake recurrence statistics for the Tashkent region

Earthquake Magnitude	Return Period (yr) Macroseismic Data	Return Period (yr) Recurrence Relationship (r = 0.46)
4.5 - 5.0	10	15
5.1 - 5.6	30	35
5.7 – 6.2	63	75
6.3 – 6.8	85	210

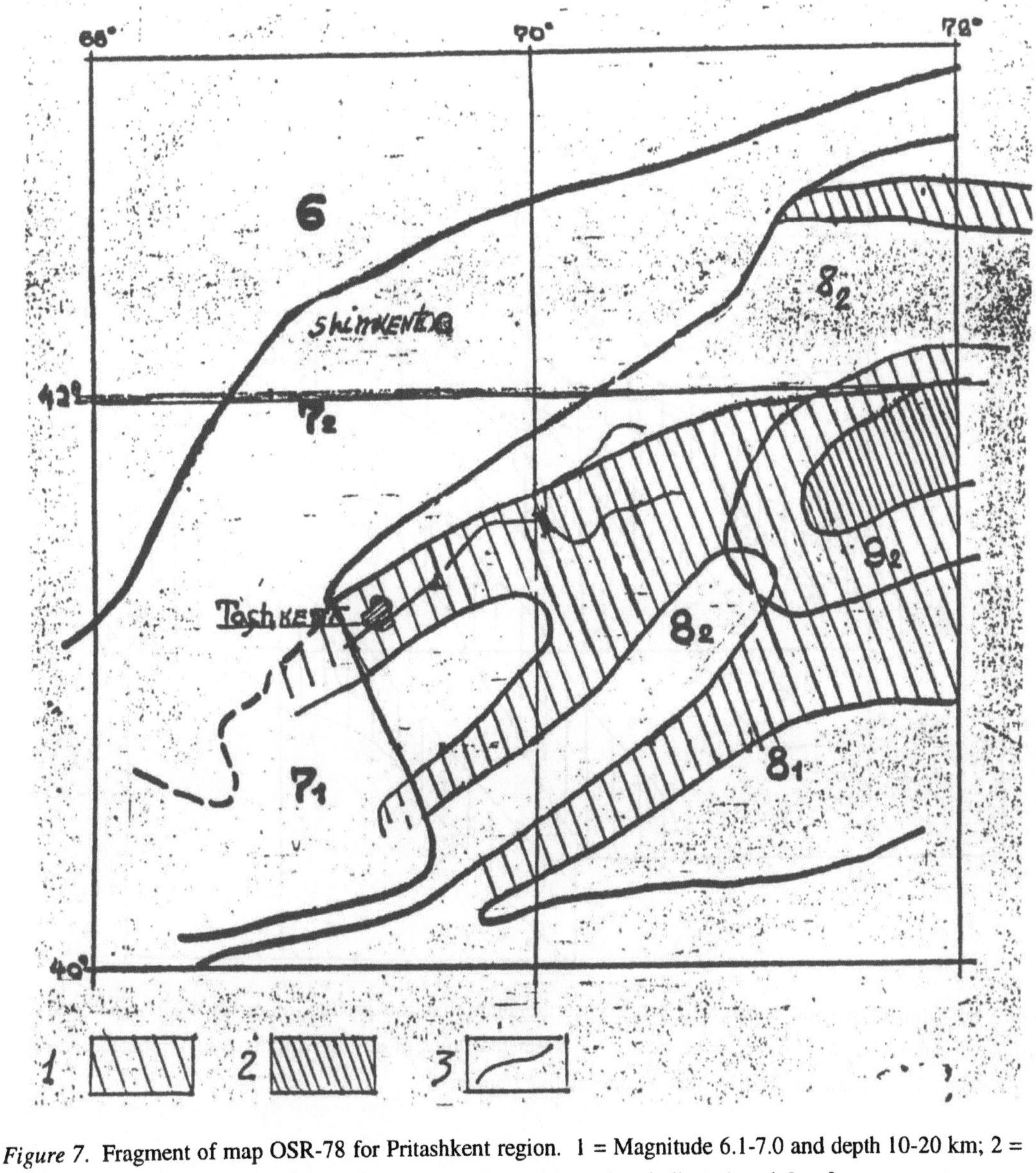

Figure 7. Fragment of map OSR-78 for Pritashkent region. 1 = Magnitude 6.1-7.0 and depth 10-20 km; 2 = Magnitude 7.1-8.0 and depth 16-30 km; 3 = isolines of intensity; indices 1 and 2 refer to average return periods of 100 and 1000 years, respectively.

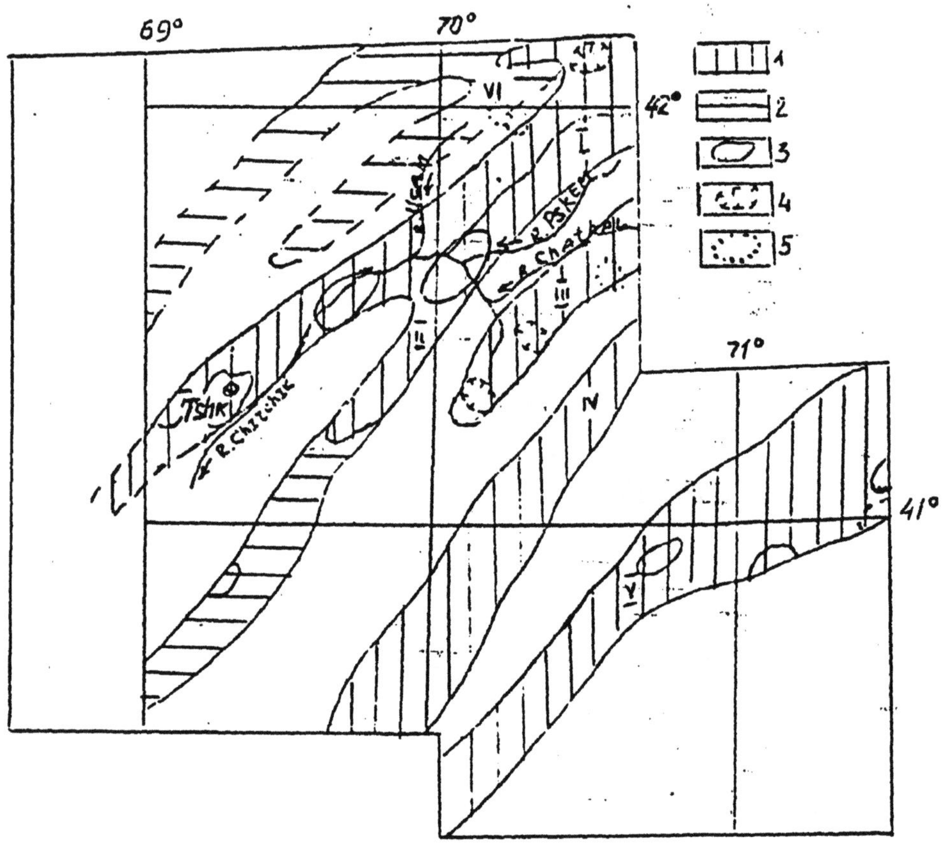

Figure 8. Map of seismic zones with possible earthquakes that can produce intensity 7 or more in the Tashkent region. 1 = M < 6.7; 2 = M < 5.5; 3 = isoseismal lines of strong earthquakes; 4 = paleoseismic dislocations; 5 = seismological dislocations from satellite imagery.

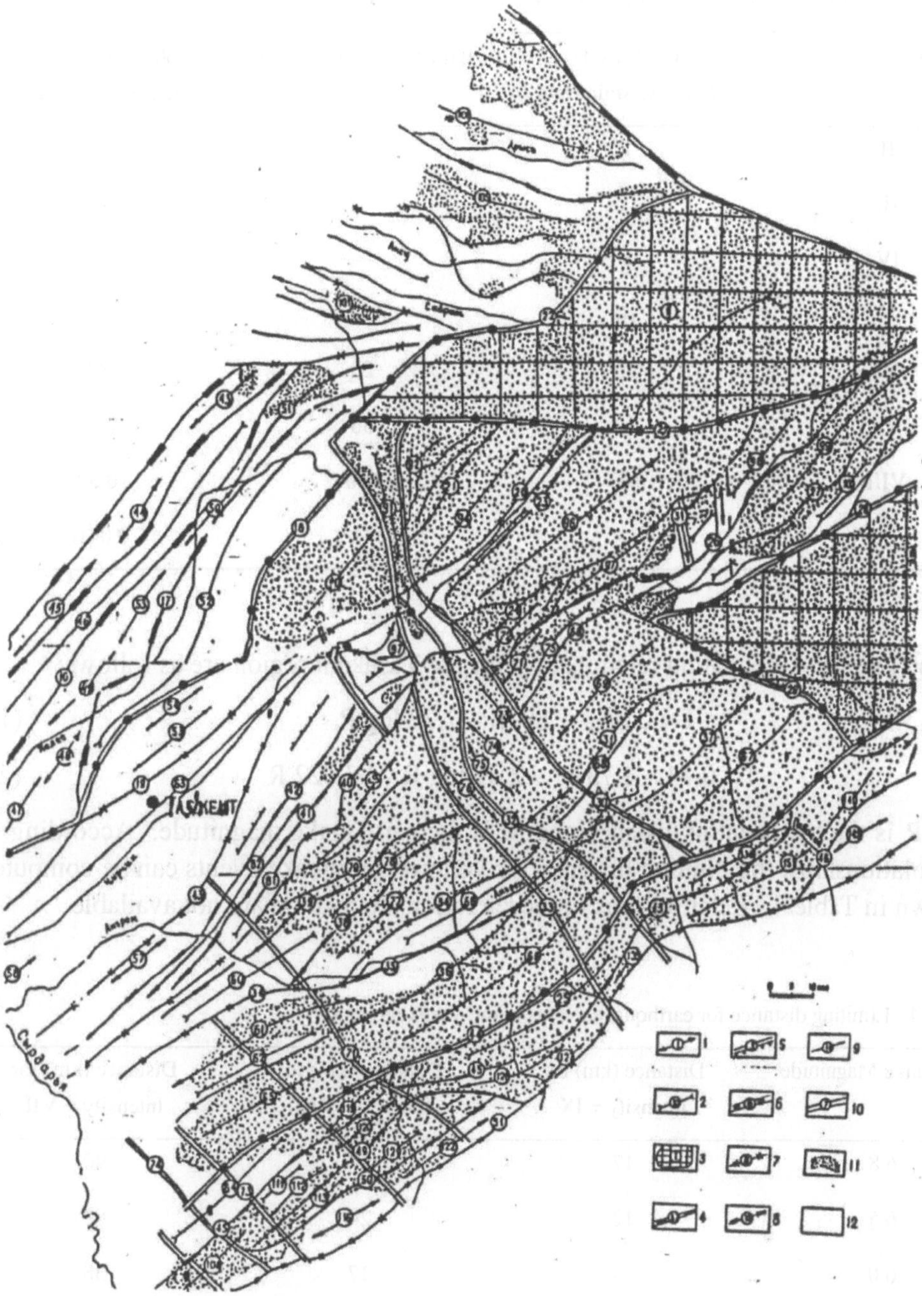

Figure 9. Schema of most recent structural forms of Middle Tien-Shan (compiled by Djamalov et al., 1990).

TABLE 3. Recurrence of seismic intensity levels for Tashkent region

Seismic Intensity	Return Period (yr) Macroseismic Data	Expected Return Period (yr)	Probability of at least one event in 50 years
II	1	---	---
III	2.5	---	---
IV	5	---	---
V	10	---	---
VI	22	20	092
VII	48	56	0.58
VIII	102	100	0.39
IX	---	1125	0.0416

The intensity attenuation relationships for the Tashkent region are as follows:

$$I = 0.24 + 1.80\,M - 2.72\log R \qquad (1)$$

$$I = 0.14 + 1.75\,M - 2.39\log R - 0.0022\,R \qquad (2)$$

where R is the hypocentral distance and M is the earthquake magnitude. According to these relationships, the limiting distance for various earthquake events can be computed as shown in Table 4. Relationships for acceleration attenuation are not available.

TABLE 4. Limiting distance for earthquake events in the Tashkent region

Earthquake Magnitude	Distance (km) for Intensity = IX	Distance (km) for Intensity = VIII	Distance (km) for Intensity = VII
6.8	17	38	90
6.5	12	24	62
6.0	---	17	38
5.8	---	12	24
5.5	---	---	17

For the Tashkent region, it is possible to develop relationships between energy class K, where K = log E, magnitude M of the earthquake, the frequency f_{disp} of maximum spectral displacement, and the frequency f_{vel} of maximum spectral velocity. The relationships are as follows for average soil conditions:

$$f_{disp} = 3.81 - 0.08\ K \quad (3)$$

$$f_{disp} = 3.49 - 0.14\ M \quad (4)$$

$$f_{vel} = 9.78 - 0.168\ K \quad (5)$$

$$f_{vel} = 9.11 - 0.30\ M \quad (6)$$

$$\log E = 1.62 + 0.46\ K \quad (7)$$

$$\log E = 2.46 + 0.83\ M \quad (8)$$

According to Ulomov et al. (1966) the expected accelerations in Tashkent for intensity levels VII, VIII, and IX are 65, 144, and 323 cm/sec^2, respectively, for a focal depth of 5 km and average soil conditions.

3. Description of Damaging Earthquakes in Uzbekistan

3.1 GENERAL INFORMATION

Information about damaging earthquakes in Uzbekistan is available dating back to the 1494 earthquake located in the north part of the North-Fergana fault, about 180 km from Tashkent. More accurate earthquake data are available since the M 6.5 Tashkent earthquake of 1868, which occurred on the Teshiktash zone and caused an intensity of IX in Tashkent. There have been 14 earthquake events of intensity VII or more in the Tashkent region during the 1868 to 1996 time period. Table 5 lists the damaging earthquakes that have occurred in the region.

Figure 10 shows a summary isoseismal map for strong earthquakes in the Tashkent region. Figure 11 shows the isoseismal map for the 1966 M 5.3 earthquake with its epicenter in the Tashkent territory. The majority of the damage due to this event occurred in the central part of the city. An area of about 10 km^2 experienced intensity VIII shaking.

3.2 DESCRIPTION OF CONSTRUCTION TYPES

In 1868, Tashkent was the administrative center of Turkestan kray, and most houses were built from local materials such as clay of the "pakhsa" type, abode bricks, and wooden frames with adobe infill. The city had primarily one-story buildings until the revolution of 1917. At the beginning of the 20th century, dwellings covered an area of about 1.4 million m^2. In the year 1932, construction was started on one to four story brick buildings for workers in Tashkent. By 1940, dwellings covered an area of about 2.8 million m^2, and about 4% of the dwellings were multistory brick buildings with wooden ceilings that were not designed according to any regulations.

TABLE 5. Summary data for strong earthquakes in the vicinity of Tashkent

Name	Date	Magnitude	Distance from Tashkent (km)	Intensity at Epicenter	Intensity in Tashkent	Fault(s)	Type of Faulting	Population (thousands)
---	2/4/1868	5-6	40	VII	VII	Tashkent-Karjantau	---	80
Tashkent 1	4/3/1868	6.5	27	VIII	VII-VIII	Toitepe (North Samsar)	---	80
Tashkent 2	11/29/1886	6.7	30	VIII	VII	Tashkent-Pskem (Karjantau)	---	156
Kostakoz	11/28/1888	6.3	145	VIII	VI-VII	North Fergana	---	156
Ura-Tyube 1	9/17/1897	6.6	170	VIII	VI-VII	---	---	106
---	7/7/1924	4.3	9	VII	VII	North Tashkent	Strike-slip	234
Chatkal	11/2/1946	7.5	230	IX-X	VII-VIII	Chatkal-Atainak	Thrust	770
Tashkent 3	4/25/1966	5.1	0	VIII	VIII	Tashkent-Karjantau	Thrust	---
Aftershock	5/5/1966	3.9	0	VII	VI-VII	Tashkent-Karjantau	Thrust	---
Aftershock	5/9/1966	4.2	0	VII	VII	Tashkent-Karjantau	Thrust	---
Aftershock	5/24/1966	3.7	0	VII	VI-VII	Tashkent-Karjantau	Thrust	---
Aftershock	6/29/1966	3.6	0	VII	VII	Tashkent-Karjantau	Thrust	---
Aftershock	3/24/1967	3.5	0	VII	VI-VII	Tashkent-Karjantau	Thrust	---
Nazarbek	12/11/1980	4.8	16	VIII	VI-VII	Karakamysh	Thrust	>1900

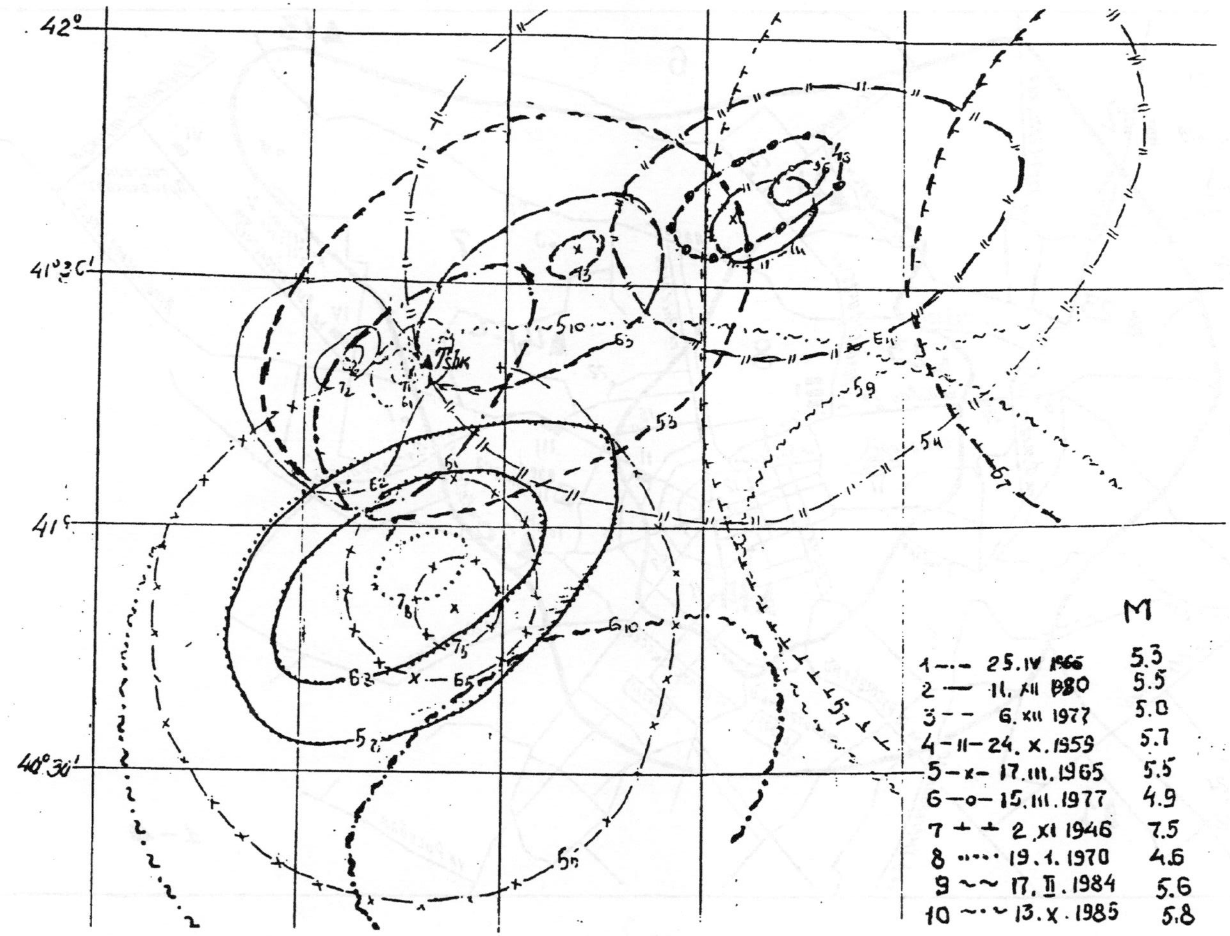

Figure 10. Map of isoseismal lines of strong earthquakes in the Tashkent region, 5.0 < M < 7.5 (Yankovskaya and Sokolova, 1989).

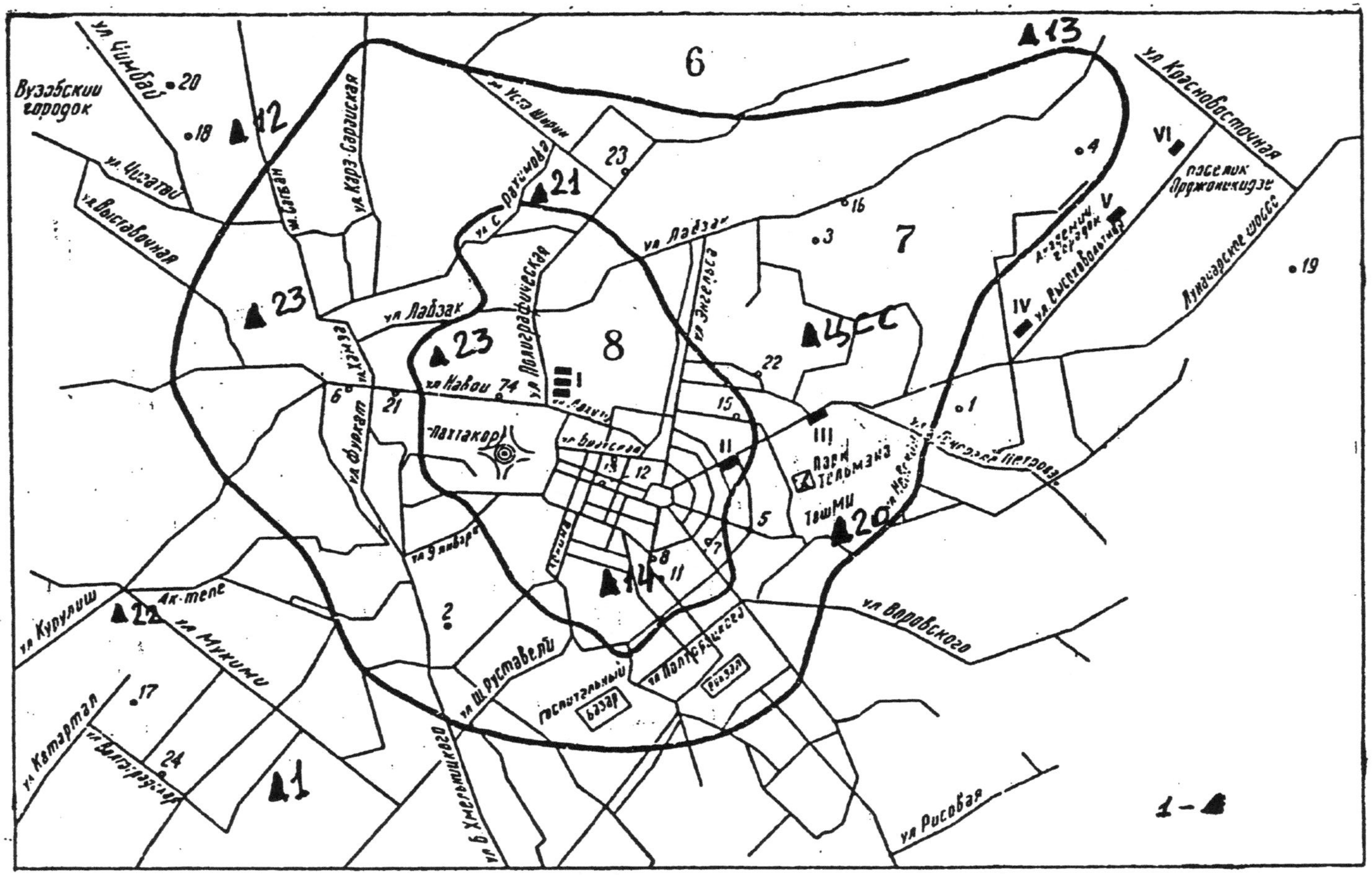

Figure 11. Map of epicenters of 1966 M = 5.3 Tashkent earthquake; 1 = Seismic stations (compiled by Rasskazovsky et al.).

In 1948, the institute "Uzgorproyekt" designed a first series of standard 2-story brick residential buildings (series N 21) with longitudinal walls spaced at 11.68 meters and transverse walls spaced at 3.9 meters. From the end of the 1940s to the end of 1957, engineers followed the technical standards TU 58-48 for design and construction of buildings for seismic regions and the standards on seismic construction PSP 101-51.

Starting in 1955, standard 4-story residential buildings of series 1-41 with brick walls, prefabricated reinforced concrete ceilings, and annexed verandas of reinforced concrete frames were constructed. These buildings have transverse walls spaced at 5.6 meters and floor heights of 3.3 meters. From 1956 to 1959, the institute "Uzgorproyekt" designed and built two to five story residential brick buildings of series 1-310. These buildings had 38 cm thick transverse walls, 6 meter bays, and longitudinal walls at 10.4 meter spacing. Ceiling heights were 2.8 meters, and ceiling joists were made of reinforced concrete. Seismic design levels corresponded to intensities of VII to VIII.

Large panel building construction began in Tashkent in 1961. The institute "Tashgiprogor" designed 4 to 5 story buildings of series TDSK to withstand earthquakes of intensity VIII. Transverse walls were spaced at 3.3 meters and ceiling heights were 4 meters. Longitudinal walls were spaced at 5.4 meters and were 12 to 16 cm thick. Ceilings spanned in two directions and were 12 cm thick. From 1963 to 1965, large panel 4-story buildings of the series 1-Uz-500 TDSK were built to a design level of intensity VII to VIII, and large panel 4-story buildings of the series 1-Uz-500 TSP were built to a design level of intensity IX. Ceiling joists in these building spanned in two directions.

At the time of the earthquake in Tashkent in 1966, the residential building space was distributed as shown in Table 6.

TABLE 6. Summary of residential building space in Tashkent in 1966

Type of Building	Total Space (x 1,000,000 m^2)
One story, houses from adobe bricks	2.970
One story, frame-adobe of fired bricks	1.880
Two story, brick	0.796
Multistory (3-5 stories) of series 1-421, 1-310, and 1-310 I	1.210
Large panel of TDSK and 1 Uz-500	0.350
TOTAL	7.206

After the 1966 earthquake in Tashkent, residential brick building of 4 to 5 stories were designed and built. These buildings were of series 1-310 I and had a design load of seismic intensity VIII. In seismic regions, buildings of series 1-310 TSP were built to an intensity level of IX. These buildings had additional structural elements, including pre-fabricated reinforced concrete frames with spans of 4.8 to 5.8 meters. From 1967 to 1970, a number of 9-story frame-panel buildings were built according to individual projects of designers from Moscow. In 1975, Institute TashZNIIEP designed 9-story large panel buildings of the 148 series, which were constructed in the early 1980s.

3.3 DESCRIPTION OF EARTHQUAKE DAMAGE

During the earthquake of 1868 in Tashkent, many buildings constructed of local materials were damaged. Corners fell and walls collapsed and many people were killed or injured. The earthquake of 1886 caused 7 buildings to collapse in Tashkent. The Russian part of the city sustained more damage than the Asian part of the city. Buildings in the Russian part of the city were primarily constructed of adobe bricks with wall height and bay width greater than the buildings in the Asian part of the city. In addition, the buildings in the Asian part of the city were typically wood frame with adobe infill walls.

Adobe buildings in Tashkent were heavily damaged in the 1966 earthquake. As a result of the earthquake, the residential building area decreased from 7.2 million m^2 to 2.8 million m^2. There were no cases of modern buildings collapsing. No buildings of local or modern construction collapsed in Tashkent during the 1980 Nazarbek earthquake.

Typical damage to buildings constructed of local materials includes separation of longitudinal walls from transverse walls, full collapse of non-load-bearing walls, collapse of corners of buildings, lamination of brick walls, collapse of pin joints, pounding of adjacent buildings of different heights, and diagonal and horizontal cracking in walls.

Typical damage to 4-story buildings in Tashkent that were built in the 1940s and 1950s includes diagonal and horizontal cracking in single pier elements, x-type cracking in bricks between apertures, and diagonal and horizontal cracking in staircases. Ten to twelve of these buildings were damaged in the 1966 earthquake (seismic intensity VIII) and were later restored. Residential multistory buildings in Tashkent are of series 1-310 and 1-310 I. The most common damage to these buildings includes diagonal and horizontal cracks in single pier elements and walls, cracks at wall-ceiling junctions, cracks in reinforced concrete ceilings, and opening of the joints between slabs and ceilings.

Frame buildings in Tashkent have not been subjected to a major earthquake. In other areas, these buildings have shown earthquake damage is possible in load-bearing structures and separation walls, including diagonal cracks in walls, cracks along the contours of the infill walls, and separation of walls and frames. Large panel building construction is fairly new and there is little information on earthquake damage to these types of buildings (series 1-Uz-500 and TDSK).

3.4 REASONS FOR EARTHQUAKE DAMAGE

The reason for extensive damage to adobe brick buildings is that they were not designed to resist earthquake shaking, and many of them were located in the epicentral zone of the damaging 1966 earthquake. In addition, many of them had experienced general deterioration before the earthquake occurred. Damage to older brick buildings is caused by factors such as complicated configurations in design, lack of seismic joints and belts, large basements under part of buildings, and irregular and asymmetric wall locations.

Modern brick buildings designed with seismic measures specified by SN-8-57 and SNiP P-A.12-62 have typically performed well in earthquakes. The main reasons for damage to these buildings include low quality of workmanship and construction materials, difference in locations of center of stiffness and center of mass, irregular configuration of load-bearing walls, large distances between walls, and variations in stiffness in longitudinal and transverse directions.

Damage to frame buildings typically occurs in partitions and infill brick walls because these elements are not properly connected to the frames.

3.5 SECONDARY EFFECTS

Secondary earthquake effects, such as landslide and liquefaction, have not been observed in Tashkent.

4. Description of Building Construction in Uzbekistan

Table 7 lists a summary of the building construction types in Tashkent. Table 8 shows the current (1996) status of the residential building construction in Tashkent. As shown in Table 8, about 43% of the inhabitants of the city live in buildings that were not adequately designed and constructed to meet the current standards for seismic resistance.

The most vulnerable building types in Tashkent include the following:

- 9, 12, and 16-story frame-panel buildings that include frame structures of the IIS-04 series – In general, many elements of these buildings are prefabricated and welded in the field. The quality of the welding is typically not good. These buildings have been constructed from 1974 to the present.
- Frame structures without diagonal bracing built by lifting the ceilings, which have a round profile and one core of stiffness – These buildings have been constructed since 1980. As the Spitak earthquake revealed, these buildings have an irregular stiffness distribution and very little reserve strength.

TABLE 7. Statistics of mass residential building construction in Tashkent (number of buildings and inhabitants)

Type of Building and Serial Number	Constructed 1951-1962 under regulations PSP 101-51 (1951-1957) SN-8-57-55 (1957-1962)	Constructed 1962-1972 under regulations SNiP P.2.12-62 (1962-1969) SNiP P-A.12-69 (1970-1977)	Constructed 1973-1985 under regulations SNiP P-A.12-69* (1977-1981) SNiP P-7-81 (1982-1990)	Constructed 1986-1996 under regulations SNiP P-7-81* (1991-1996) KMK 2 01.03-96 (1996-pres.)
Brick w/ longitudinal load-bearing walls (series 1-421)	340 ; 19,000	---	---	---
Brick w/ transverse load-bearing walls (series 1-310)	960 ; 51,000	1370 ; 72,000	1370 ; 72,000	---
Brick construction w/ transverse load-bearing walls (series 77)	---	410 ; 22,000	3960 ; 210,000	860 ; 45,000
Reinforced concrete frame w/ brick walls (series 1-310 TSP)	270 ; 14,000	860 ; 45,000	---	---
Reinforced concrete frame-panel	---	---	1130 ; 64,000	1000 ; 56,000
Reinforced concrete large panel (series IUz-500 TSP-1, 148, 148P, 1-T-SP, P-TSP)	182 ; 10,000	2860 ; 147,000	8480 ; 439,000	3878 ; 200,000
Monolithic reinforced concrete	---	---	150 ; 7800	100 ; 5200
9-story residential buildings from 3D blocks (series BTS)	---	---	150 ; 7500	350 ; 17,500
1-story houses of adobe, brick, wood frame w/ adobe, etc.	---	6830 ; 22,000	---	---

- Brick residential buildings built before 1966 and of series 1-310 built 1954 to 1962 – These buildings do not have interior longitudinal walls or reinforced concrete cores. In addition, many have experienced damage due to foundation settlement.
- Brick buildings of series 1-310 I built after 1966 – These buildings often have low quality of workmanship and poor quality of construction materials. There are no means for controlling processes such as setting of bricks, vibration of concrete, and filling of joints with mortar.

TABLE 8. Characteristics of residential building construction in Tashkent

	Total Buildings		Buildings Not Corresponding to Seismic Resistance Requirements		Relative Seismic Resistance
Structural System	Area in million m^2 (% of total)	Inhabitants in thousands (% of total)	Area in million m^2 (% of total)	Inhabitants in thousands (% of total)	Scale of 1-100
3D blocks of reinforced concrete	0.50 (1.4)	25 (1.2)	0.01 (0.03)	0.5 (0.02)	98
Large panel reinforced concrete	15.4 (43.4)	796 (38)	0.77 (2.17)	40 (1.69)	95
Monolithic reinforced concrete	0.25 (0.7)	13 (0.6)	0.07 (0.11)	2.01 (0.1)	84
Brick	10.4 (29)	550 (26.2)	4.96 (14)	262 (12.5)	52
Frame-panel reinforced concrete	2.13 (6)	120 (6)	1.23 (3.46)	69 (3.3)	42
1-2 story houses (adobe brick, wood frame with adobe infill, etc.)	6.83 (19.5)	584 (28)	6.28 (17.7)	532.5 (25.5)	8
TOTAL	35.5 (100)	2088 (100)	13.29 (37.4)	906 (43.3)	

In Tashkent, the area of multistory (4 or more stories) buildings that are of the dangerous construction types is estimated at approximately 1,750,000 m^2, with 217,000 inhabitants.

5. Research Institutes and Organizations in Uzbekistan

The institute that develops standards for seismic design and construction in Uzbekistan is the Uzbek Research and Design Institute of Standard and Experimental Design of Residential and Public Buildings (Joint Stock Society UzLITTI. 17, Niyazov St., Tashkent, 700095, phone: 46-74-66, 46-07-03). Additional participation is provided by the following:

- Institute of Seismology AS RU - 3, Khurshid St., Tashkent, 700128, phone: 41-51-70
- Institute of Mechanics and Seismic Resistant Construction AS RU (IMiSS) - 143, Akademgorodok, phone: 62-71-32
- Tashkent Architecture-Construction Institute - 13, Navoi, Tashkent, phone: 41-15-01, 41-80-02
- Geology and Geophysics Institute AS RU - 41, Khodjibaeva St., Tashkent, phone: 62-65-16, 62-68-95

The regulatory documents undergo review and are then approved and mandated by the order of the State Committee of the Uzbek Republic on Architecture and Construction (Abay St., Goskomarkhitekurstroy RU, phone: 44-07-00, 39-86-96).

Users of the standard documents are all institutes of construction projects, the largest of which are the following:

- UzNIIPGradostroitelstva – 18, Navoi St., Tashkent, 700011, 41-44-64, 41-45-92
- Tashgiprogor – 40, Navoi St., Tashkent, 42-25-97, 42-22-87
- Uzgiproselstroy – "TS" kvartal, Volgogradskaya St., Tashkent, 77-12-43, 77-76-76
- AO "Uztayazhprom" – 88, Pushkin St., Tashkent, 33-99-84, 68-85-80
- TashNIPI of general planning – 14, Babur St., Tashkent, 55-50-43, 55-65-49

Mass construction in Tashkent is carried out by the corporation "Tashinveststroi" (formerly Glavtashkentstroi) which was organized in 1963 (26a, Uzbekistansky Prospect, Tashkent, 33-90-33, 45-44-42). More that 90% of the total volume of construction of civil buildings is done by this company. The following two companies also construct residential buildings in Tashkent:

- "Uzpromgrazhdanstroi" Corporation (formerly Ministry of Construction) – 17, Movarounnakhr St., Tashkent, 33-77-25, 36-01-04

- "Uzvodstroi" Concern – 6, Abai St., Tashkent, 44-04-61

6. Regulations for Seismic Resistant Design and Construction in Uzbekistan

Construction norms and regulations that have been in force in Uzbekistan include the following:

- PSP 101-51: "Regulation on Construction in Seismic Regions," Gosstroyisdat, 1951
- SN-8-57: "Norms and Regulations for Construction in Seismic Regions," Gosstroi of the USSR, in operation until 1962
- SNiP P-A, 12-62: "Construction in Seismic Regions, Design Norms," Gosstroi of the USSR – Moscow, Stroyizdat, 1963, in operation from 1963 to 1970
- SNiP P-A, 12-62: "Construction in Seismic Regions, Design Norms," Gosstroi of the USSR – Moscow, Stroyizdat, 1970, in operation from 1970 to 1977
- SNiP P-A, 12-69*: "Construction in Seismic Regions, Design Norms," Gosstroi of the USSR – Moscow, Stroyizdat, 1977, in operation from 1977 to 1982
- SNiP P-A, 12-81: "Construction in Seismic Regions, Design Norms," Gosstroi of the USSR – Moscow, Stroyizdat, 1982, in operation from 1982 to 1991
- SNiP P-A, 12-81*: "Construction in Seismic Regions, Design Norms," Gosstroi of the USSR – Moscow, Stroyizdat, 1991, in operation from 1991 to 1996
- KMK 2.01.02-96: "Construction in Seismic Regions, Design Norms," Goskomarchitectstroi RUZ – Tashkent, 1996, valid from 1996 to the present

6.1 THEORETICAL BASIS FOR REGULATIONS

In the standards PSP 101-51, which were in operation from 1951 to 1957, design and construction of buildings were based on static theory according to methods developed by F. Omori of Japan. The seismic load was defined according to the following formula:

$$S = K_c K_c^* Q \tag{1}$$

where K_c is the seismic coefficient with values 0.1, 0.05, and 0.25; K_c^* is the soil factor with values of 0.025 to 0.5, and Q is the weight of the building. This method did not give the actual distribution of the earthquake loads within the building.

In the document SN-8-57, the seismic design methods included the use of a spectral function based on the period of maximum acceleration. Calculation of possible loads that correspond to different forms of free oscillations was made with the following formula:

$$S_K = Q_K\, K_c\, \beta\, h\, k \qquad (2)$$

where K_c is the seismic coefficient with values 0.025, 0, and 0.1; Q_K is the load-causing inertial force equal to 1 for constant loads and 0.8 for temporary loads; β is the dynamic coefficient varying from 0.6 to 3 and equal to 0.9/T, where T is the modal period of the structure; and k and h are coefficients depending on the mode shape.

In SNiP P-A 12-62, the design formula and parameters remained the same; however, they were corrected to account for load type values and load factors. In SNiP P-A 12-69, the design formula and parameters also remained the same; however, the value of the dynamic coefficient changed to $1/T_i$ with a range of 0.8 to 3.0. In addition, the seismicity of the site was corrected to account for local soil conditions. Load requirements were divided according to the type of building (e.g., frame, large panel, and brick). Also, the load combination coefficient was changed to 0.9 for constant loads, 0.8 for temporary loads, and 0.5 for roof and snow loads. SNiP P-A 12-69* included a few changes in the treatment of soil categories and the requirements for construction methods. In addition, use was made of the new USSR seismic zoning map.

In SNiP P-7-81, the formula for defining the seismic load was changed as follows:

$$S_{ik} = K1\; K2\; Q_k\; A\; \beta_i\; K_\psi\; h_{ik} \qquad (3)$$

where K1 is a coefficient equal to 0.25 that accounts for damage; K2 is a coefficient for the building type that ranges from 0.5 to 1.5; Q_k is the weight of the building to point "k;" A is a coefficient for the seismic intensity equal to 0.1, 0.2, or 0.4 for intensity VII, VIII, and IX, respectively; K_ψ is a coefficient from 1 to 1.5 to account for irregular construction; and β_i is the dynamic coefficient equal to 0.8/3 for soil category I, 0.8/2.7 for soil category II, and 0.8/2.0 for soil category III. This document also included requirements for interior longitudinal walls in buildings with brick load bearing walls.

SNiP P-7-81* included changes to account for the new seismic zoning maps. For sites with intensity greater than IX, construction was allowed only with approval from the government. Changes were also made with respect to the definition of soil categories.

The KMK 2.01.02-96 document included several changes to the design and construction regulations as follows:

- The codes were obligatory for buildings being designed and built in seismic zones with intensity VII or higher
- Maps of seismic zoning and information on earthquake recurrence were updated
- Classification of soil types and the increase in intensity for various soil types were improved
- Two methods of design were specified – one for the physical design and one using the elastic response spectrum

- New parameters were included to account for seismicity in regions with intensity greater than IX
- New parameters were included to account for structural response, earthquake recurrence, and plan and height irregularity
- Construction requirements were made for individual houses with local materials, monolithic frame buildings, concrete hollow block buildings, and architectural elements
- Requirements for reinforcing concrete were improved
- Seismic resistant design of underground structures, repair and rehabilitation of buildings, and quality control of workmanship issues were addressed for the first time

The design formula for defining the seismic load according to the spectral method was specified as follows:

$$S_{ik} = K_{res}\, K_{rep}\, K_{fl}\, K_{reg}\, S_{oik} \quad (4)$$

where S_{oik} is the inertial force assuming linear structural behavior calculated with the following formula:

$$S_{oik} = Q_k\, W_i\, K_d\, h_{ik} \quad (5)$$

where Q_k is the weight of the building up to point "k;" W_i is the spectral coefficient; K_d is the coefficient of dissipation; h_{ik} is a coefficient depending on the mode shape; K_{reg} is a coefficient of regularity; K_{res} is a coefficient of building response (0.8 to 1.5); K_{fl} is a coefficient depending on the number of floors; and K_{rep} is a coefficient for the recurrence of earthquakes (0.8 to 1.25).

6.2 ENFORCEMENT OF DESIGN REGULATIONS

Assessment of compliance with the design regulations is made by the organization Glavgosekpertiza of Goskomarkhitektura RU before the project has started and after its completion. In addition, supervision of the construction process to ensure seismic resistant considerations are met is made by members of the State Architecture-Construction Control Organization.

7. Seismic Strengthening of Existing Buildings in Uzbekistan

7.1 METHODS FOR SEISMIC STRENGTHENING OF BUILDINGS

In conjunction with major earthquakes, beginning with the 1966 event, design institutes of Uzbekistan have developed different methods for strengthening of buildings. In the republic, a great deal of experience on designing and strengthening earthquake damaged buildings has been accumulated. However, strengthening of existing residential buildings has not been successful because of their insufficient seismic resistance. There have been experimental projects on upgrade and seismic strengthening of 2 to 4-story residential buildings of series 1-210 and 1-310 and 4-story large panel buildings of series 1-164 (developed by UzLITTI, formerly TashZNIIEP), but the projects have not been implemented. The institute TashZNIIEP developed "Recommendations on Modernization, Reconstruction, and Anti-seismic Reinforcement of Residential Buildings" in 1988.

The main methods for rehabilitation of brick and rock buildings in Uzbekistan include:

- Gunite covering of walls with layers of high quality concrete mortar over steel reinforcement meshes
- Adding steel and reinforced concrete rings to narrow single-pier elements
- Adding prestressed strengthening structures of reinforced concrete or steel
- Adding special joints for shear, tension, and torsional strength

The following two methods are recommended for reinforcing large panel buildings:

- Strengthening reinforced concrete joints by adding polymer mortar based on research led by TashZNIIEP and TbilZNIIEP, and experience of increasing the seismic resistance by rehabilitation and strengthening large panel residential buildings in Gazli (Bukhara district)
- Strengthening by adding prestressed steel belts

For strengthening of frame systems the following methods are used:

- Injection of cracks with polymer mortars
- Adding vertical and horizontal steel ties
- Adding reinforced concrete or steel rings to columns and beams
- Strengthening of ceilings by adding reinforced concrete

When developing building repair or rehabilitation projects, it is often necessary to consider the current state of the building in the design of the project. The following methods are recommended for this purpose:

- When not changing the existing construction scheme of the building, rehabilitation should include strengthening of existing elements and the ties between them
- When partially changing the existing construction scheme of the building, improvements should be made to the spatial layout and the seismic resistance of the building by adding relatively stiff longitudinal and transverse walls and seismic joints

- When totally changing the existing construction scheme of the building and its dynamic properties, improvements include adding stiff diaphragms and ties between them

7.2 ORGANIZATIONS WORKING ON SEISMIC STRENGTHENING

All of the organizations listed in Section 5 that work on building design and construction also work on seismic strengthening of buildings. The specialized institute Tashzhilproject (43 Dubitsky St., Tashkent) also works in this area. Building repair and rehabilitation are also performed by the organization Tasgzhilremont.

There are no specialized organizations that deal only with seismic strengthening of buildings. Requirements for strengthening buildings were first described in the document KMK 2.01.03-96, "Norms and Regulations for Construction in Seismic Zones." Building designers generally follow these requirements when developing projects for increasing the seismic capacity of buildings.

7.3 OBSTACLES TO SEISMIC STRENGTHENING

The primary obstacle to seismic strengthening of buildings in Uzbekistan is the lack of standard and government regulated procedures for addressing all the issues associated with increasing the seismic resistance of deficient buildings.

7.4 DESIGN LEVELS FOR SEISMIC STRENGTHENING

Seismic strengthening should bring the design level of the building up to the current requirements described in the document KMK 2.01.03-96, "Norms and Regulations for Construction in Seismic Zones."

8. New Approaches to Seismic Design in Uzbekistan

New projects that deal with seismic resistant design of residential buildings are being carried out by UzLITTI, Tashgiprogor, UzNIIPgradostroitelstva, Uzgiproselstroi, TashNIPIgenplan, Uztyazhprom, and other organizations. Requirements for seismic resistance are provided in the document KMK 2.01.03-96, "Norms and Regulations for Construction in Seismic Zones." In all cases, the design seismic load should correspond to the maximum seismic intensity expected at the construction site. Other documents that describe climatological factors, loads and impacts, and beddings and foundations are also used depending on the building materials and other equipment.

Outdated design documents are seldom used for new building construction. However, residential and public bearing frame buildings of the old series IIS-04 are still being constructed.

The following steps should be followed to increase the seismic reliability of residential buildings:

- Strictly follow the code requirements during design with supervision by Glavgosekspertiza
- Construct the building according to the approved design plans with supervision by Gosarkhstroikontrol
- Use high quality materials and structural elements in the construction with supervision by Gosarkhstroikontrol
- When possible, implement modern design procedures through the use of computers, utilize new methods of seismic protection, and obtain actual site information about structural properties, ground conditions, and seismicity

Problems with achieving adequate seismic reliability of residential buildings include the following:

- Almost total lack of local and foreign information about the latest achievements in the fields of seismic resistant construction and engineering seismology
- Lack of additional data on effective legislation to stimulate designing of earthquake resistant buildings and the strengthening of existing hazardous buildings (e.g., insurance incentives)
- Lack of continuous training, workshops, and seminars for improving the technical capabilities of specialists in various fields
- Lack of training materials, for specialists and the general public, on measures for increasing the seismic resistance of residential buildings

9. Optimal Methods for Defining Seismic Loads in Uzbekistan

The information provided to building designers by seismologists is in the form of seismic zoning maps, which are not adequate. What is required is a database containing local information such as actual recordings of acceleration, velocity, and displacement time histories and hydrogeological ground conditions. Actual or synthetic accelerograms would be much more useful than seismic intensity levels. Recurrence intervals for intensity and peak accelerations based on historical data would also be useful.

There is a service for recording earthquake shaking on the ground and in buildings at the Institutes of Seismology of AS RU and IMISS AS RU. The peak accelerations that have been recorded by this service are 110 to 150 cm/sec^2 for an event with intensity VIII to IX.

10. Scientific and Technical Cooperation for Reducing Seismic Risk

Possible forms of collaboration include the following:

- Seismic resistance of underground utilities in cities; standards for calculating seismic loads on buildings; methods for seismic strengthening of buildings; methods for reducing damage from future earthquake events
- Considering the specific problems in various settlements, development of methods for building design, construction control, and training of citizens on earthquake preparedness
- Conservation, restoration, and seismic strengthening of architectural and historical monuments in hazardous regions
- Development of effective methods for inspecting construction projects, including criteria for adequate seismic safety
- Accurate evaluation of seismic hazard in various towns and settlements and development of methods for reducing damage in future earthquakes
- Assessment of technological factors that can change the estimated seismic hazard and earthquake recurrence in various developed regions
- Development of effective methods for improving the seismic resistance of brick and masonry block buildings, including adding adhesive mortars and non-metallic reinforcement
- Development of methods for restoration and seismic strengthening of buildings constructed of low quality materials (strength less than 7.5 MPa)
- Creation of training and educational films on earthquake protection of buildings and the general public
- Introduction of new taxes, insurance, or other legislative acts as effective measures to reduce seismic damage potential
- Development of research programs on building design and construction for static and dynamic loading cases
- Creation of a database that includes behavior of various buildings types during earthquakes, and technical, economic, and damage potential characteristics of various building types
- Development of effective and automated methods for design and monitoring of large rockfill dams
- Experimental work on modern dynamic testing of model buildings, including nondestructive techniques
- Development of seismic resistant designs for basements and foundations in areas with poor soil conditions

- Development of methods for active control of buildings when subjected to earthquake shaking
- Creation of regional training centers on issues and problems related to seismic resistant design and construction

DESTRUCTION OF STANDARD RESIDENTIAL BUILDINGS IN THE 1988 SPITAK, ARMENIA EARTHQUAKE

T. MARKARIAN
Ministry of Urban Development, Armenia
S. SHAGINIAN
Ministry of Urban Development, Armenia

1. Introduction

There are numerous reports by specialists of many countries dedicated to the Spitak, Armenia earthquake of December 7, 1988. These reports give analyses of the main causes and characteristics of damage to buildings of different designs, types, and heights, as well as the features of the earthquake shaking on various geological conditions. This chapter has been prepared on the basis of results of inspections of residential and public buildings in the regions damaged by the 1988 Spitak, Armenia earthquake made by specialists from research and design institutions in Armenia, and on the basis of the large amount of published materials on scientific and technical aspects of the earthquake. Material included in this chapter was also taken from the findings of a state commission of the former USSR on inspection of the quality of mass design and construction of buildings in the northern regions of Armenia.

2. Background on the 1988 Spitak Earthquake

On December 7, 1988 at 11:41 AM local time, a large earthquake occurred in the northern region of Armenia causing numerous casualties and massive destruction of buildings and other structures. It was the most devastating earthquake in the territory of the former USSR since the 1948 Ashgabad earthquake. The epicenter was located at 40.88° north latitude and 44.29° east longitude. The earthquake magnitude was 7.0 and it occurred at a depth of 15 to 20 km. Near the town of Spitak, the earthquake rupture of approximately 13 km was observed on the ground surface. In the epicentral region, near the village of Nalband, the intensity of shaking was determined to be more than X according to the MSK-64 scale. The main shock was followed by several aftershocks, the largest of which was a magnitude 5.9 event that occurred four minutes later in the region 6 to 7 km south of the main shock.

The earthquake affected 40% of the territory of Armenia, which had a population of about one million people. The earthquake severely damaged 21 towns and 365 villages,

S.A. King et al. (eds.),
Seismic Hazard and Building Vulnerability in Post-Soviet Central Asian Republics, 181–196.

58 of which were practically destroyed. The areas of heavy damage included Leninakan (70% destroyed); Spitak (95% destroyed), Kirovokan (5% destroyed), and Stepanovan (5% destroyed). The death toll exceeded 25,000 people. About 24,000 people had injuries, of which approximately 12,000 were hospitalized. More than 500,000 people were left homeless as a result of the earthquake. Roughly 7,500 apartment building covering approximately 8,000,000 m^2 in towns and villages were damaged. In rural areas, damage and destruction was observed in about 54,000 dwellings, 2,400 cattle farms, 277 schools and kindergartens, 250 public health offices, and 324 club houses and palaces of culture. Total loss to the national economy was estimated at about 10,000,000 Rubles.

Mass construction of buildings in towns and large villages that were destroyed by the earthquake included standard 4- and 5-story brick apartment buildings of series 1-451 and 1A-450, and standard 5, 9, and 12-story frame-panel buildings of the 111 series.

The town of Leninakan (now named Gumri) is the second largest in the Republic of Armenia. It is located in the center of the Shirak industrial region and covers an area of approximately 36 km^2 with a population of roughly 220,000 people. Since the end of the 1950s, construction of standard multi-story apartment buildings had been carried out in Leninakan. These buildings were built according to the following standard designs:

- Stone buildings up to 5 stories of series 1-451 and 1A-450
- Frame-panel buildings of 5, 9, and 12 stories of series 111
- Large panel buildings of 9 stories of series A1-451 KP-16/1

There were more than 12,000 apartment buildings in Leninakan, about 11,000 of which were constructed of brick. One 10-story building and one 16-story building were constructed by the lift slab method.

The following sections of this chapter give general information on the design and construction of the various types of standard multi-story apartment buildings that were built in Leninakan. These buildings were also constructed in towns and villages throughout the entire territory of Armenia.

3. Frame-Panel Buildings of 111 Series

The 111 series buildings (see Figure 1) are single-section buildings of dimensions 18 meters by 18 meters, and multi-section buildings composed of sections 18 meters by 12 meters. In the longitudinal direction they have frames with load-bearing beams, and in the transverse direction they have frames with rigid diaphragms and panels. The columns are spaced in the longitudinal direction at 4.5 meters and in the transverse direction at 6.0 meters for the multi-section buildings, and at 6.0 meters in both directions for the single-section buildings.

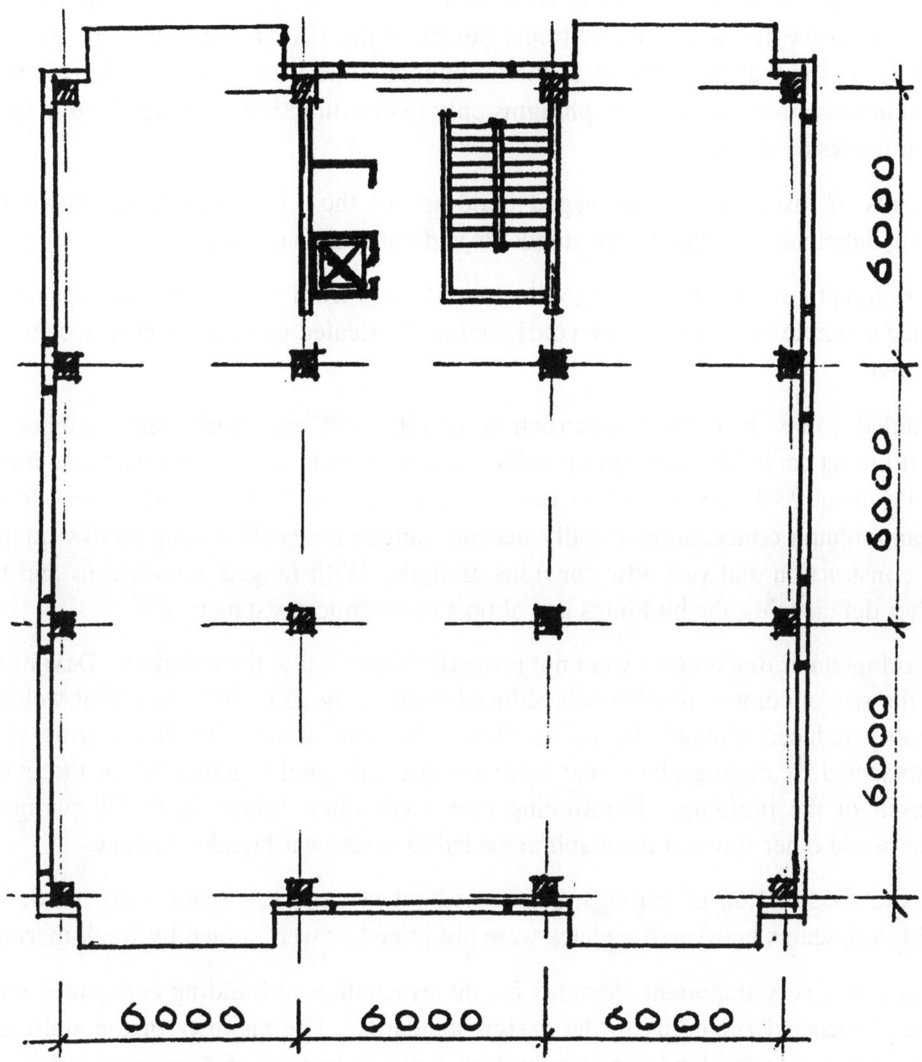

Figure 1. 9-story casing-panel residential house of series 111.

The design and construction of the frame elements and connections is similar to the standards of the frame buildings of series IIS-04, which were designed for seismic regions of the USSR. In this design, frame modules with separations in linear elements have columns with ferroconcrete joints and welded beam-column connections. The joints in the zones of maximum tension appeared to be most vulnerable and were not able to sustain ductile deformations or have any reserve load bearing capacity when earthquake shaking caused tensile forces larger than the design values. Seismic resistance of the buildings is totally dependent on the quality of construction of the connections in the frames. An additional defect of the IIS-04 frame type of the 111 series buildings is that the stiffness in the longitudinal and transverse directions is very different because there are rigid diaphragms only in one direction. This resulted in large torsional modes of vibration.

Results of inspections of damaged buildings of the 111 series have shown the following deficiencies in the design standards and construction process:

- The strength of concrete in the joints was different from that in the frames, cast-in-place concrete contained many voids, and prefabricated elements had low concrete strength.
- Welded joints had poor construction quality and an insufficient number of reinforcing bars. The strength of welded connections in reinforcing bars was found to be about 23% less than expected due to design and construction flaws. Rigid beam-column connections actually became hinged connections due to low quality of construction and very low concrete strength. With hinged connections and the other deficiencies, the buildings had almost no seismic resistance.
- Rigid modular diaphragms were not properly connected to the columns. Deviations in the size of connecting elements reduced strength by 25 to 40%, and poor welding quality reduced strength by 35 to 50%. In some cases, the diaphragms were intersected by ceiling plates and were not properly tied together along the entire height of the building. Reinforcing bars were often folded in the diaphragms. These and other flaws in the diaphragms led to severe earthquake damage.
- In the construction of ceilings, plates without seismic resistance were used. In addition, seams between the plates were not joined or were joined by weak mortar.
- Staircases, very important elements for the evacuation of building occupants, were not constructed according to the design standards. The majority of the staircases were welded to the landings at only 2 or 3 places instead of 4 as specified in the design, and in some cases they were not welded at all. In some of the connections, weak straps and deteriorated welding materials were used, resulting in strength reduction by a factor of 2 to 3. During the earthquake, there was mass collapse of staircases in buildings in Leninakan.

The points listed above, which relate to both design deficiencies and low construction quality, undoubtedly led to the increase in deformations in frames and connection elements during the earthquake. The increased deformations together with the characteristics of the earthquake shaking in Leninakan, such as the intensity and duration of strong shaking, the simultaneous horizontal and vertical shaking, and the frequency content of the motion coinciding with the natural frequencies of buildings, led to mass destruction and collapse of 9-story buildings of the 111 series. Among the 138 frame-panel 9- to 12-story buildings constructed in Leninakan, 95 were completely destroyed and 43 were so heavily damaged that they had to be torn down.

Using ground motions recorded in Gukasyan, theoretical analysis of the seismic behavior of series 111 buildings in the Spitak earthquake has shown that the seismic resistance of the 9-story frame-panel buildings is not sufficient to resist shaking of intensity IX or more. The maximum intensity that these buildings can withstand, assuming that they are constructed according to the design standards, is VII to VIII.

4. Buildings of 1A-450 Series with Brick Load-Bearing Walls

Buildings of this series (see Figures 2 and 3) are comprised of sections of serial, edge, and block-house types of structures. The serial type design scheme includes transverse load-bearing walls with an exterior longitudinal wall replaced by a ferroconcrete frame. The edge section design scheme includes 6 meter spans with transverse load-bearing walls, and 3 meter and 7.5 meter spans with longitudinal load-bearing walls. An exterior wall is replaced by a ferroconcrete frame. The block-house design scheme is composed of 4 transverse and longitudinal load-bearing walls as shown in Figure 3. Exterior walls have wide openings of 5 meters in width that are supported by ferroconcrete frames. Two middle spans with widths of 3.1 meters are also replaced by frames. Earthquake damage to these building is shown in Figures 6 and 7.

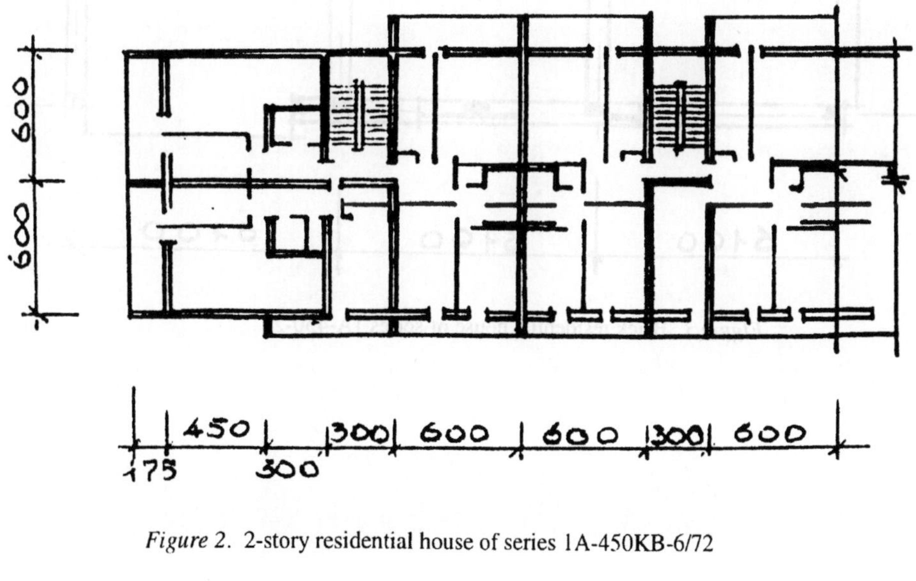

Figure 2. 2-story residential house of series 1A-450KB-6/72

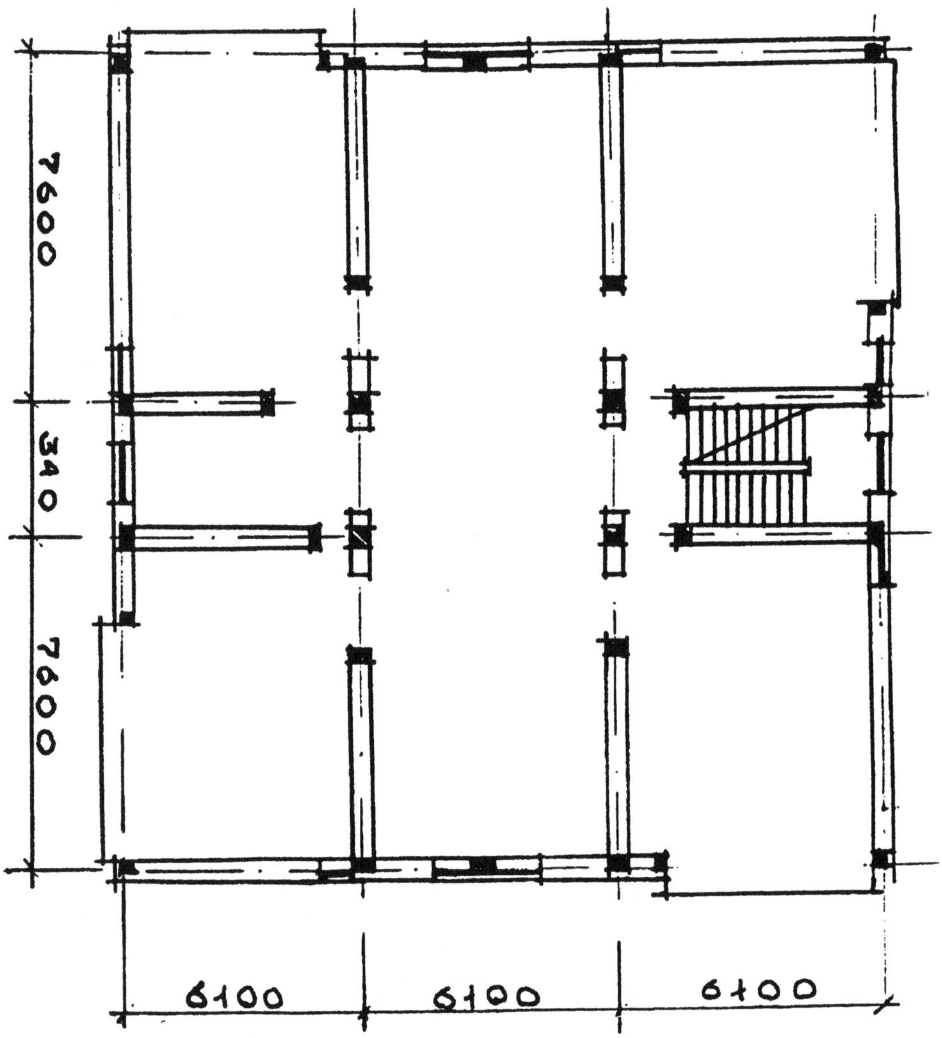

Figure 3. Brick residential house of series 1A-450-20

A modification to this series is the standard 5-story building of concrete blocks of series IA-450 KB. The majority of the design is identical to that described above for the 1A-450 series. Transverse walls are designed of large concrete blocks with height equivalent to a typical store. Connection of wall blocks is made with cast-in-place concrete reinforced with 12-14 mm diameter reinforcing hinges.

5. Buildings of Series I-451 with Three Load-bearing Longitudinal Walls

Series I-451 buildings (see Figure 4) were designed in 1958 and used for public construction in several areas of Armenia, in particular Leninakan. The design was considered the most typical of the first generation of standard construction. It had many modifications, but generally was used for apartment buildings of 4 to 5 stories located in regions with maximum expected seismic intensity of VII to VIII. The longitudinal load-bearing walls in these buildings were made from "Midis" type bricks 50 cm thick, or regularly shaped stones 40 cm thick. The buildings also included transverse walls connected to the longitudinal walls with seismic belts of monolithic reinforced concrete.

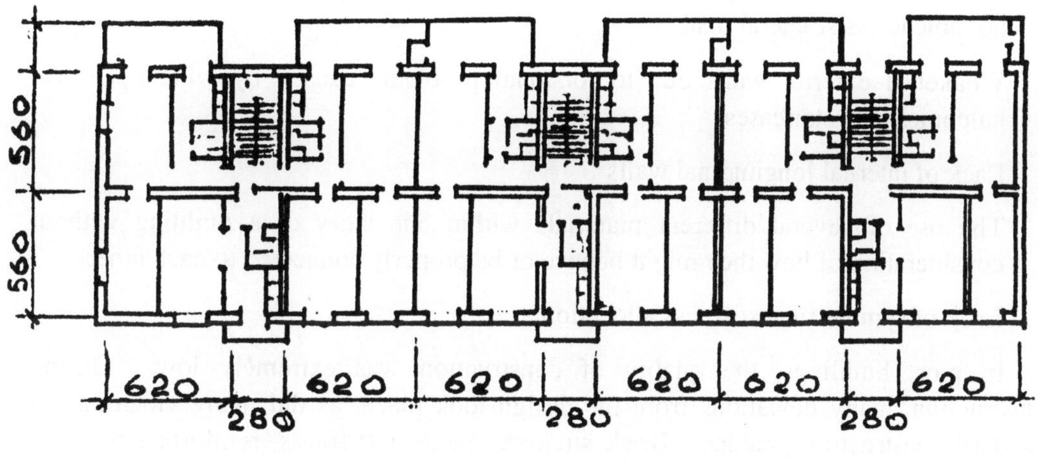

Figure 4. 5-story residential house of series I-451-I3

Analysis of the performance of stone buildings in Leninakan after the Spitak earthquake showed that the damage was highly dependent on the construction method, wall materials, and other variations in design. In particular, series I-451 4-story buildings made completely of stone with symmetrically placed walls were severely damaged, but did not collapse (see Figures 8 and 9). In these buildings, non-load-bearing exterior walls in the upper stories typically sustained the most damage due to the lack of proper connections to the longitudinal walls and the failure of continuity in the monolithic reinforced concrete seismic belts. Damage to series I-451 buildings with

"Midis" type brick walls included exfoliation of brick, opening of connections made of reinforced concrete, displacement of reinforced concrete staircases, and opening of vertical junctions between large wall blocks (in buildings with series suffix KB). Earthquake damage to buildings with walls of complex construction was much more severe.

Many buildings were completely destroyed, or damaged so heavily that restoration was very difficult. Buildings with a soft first story were severely damaged. In some cases the buildings collapsed at the first story and were shortened by one floor. In the majority of the buildings, wide deep cracks occurred in the walls, especially at locations where bricks were connected to concrete elements. In the exterior sections of many buildings, staircases were destroyed.

The most important aspects to ensuring the seismic resistance of a building are the quality of the construction and how well the design standards are followed. In Leninakan, these aspects were not present, as many design standards were flawed and almost all buildings had errors in their design and deficiencies in the quality of construction. Design and construction problems that contributed to the earthquake damage include:

- Lack of stiffness in the ground floor of buildings resulting in a soft first story that is not able to resist lateral loads
- Weakened exterior walls due to torsional problems caused by wide openings, balconies, and staircases
- Lack of internal longitudinal walls
- The use of several different materials within one story of a building without consideration of how they might behave or be properly connected to each other
- Lack of symmetrical stiffness allocation

In most buildings, the quality of construction was extremely low. During construction, many deviations from the design took place, as did many violations of standard construction practice. Brick strength, element stiffness, reinforced concrete strength, and connections between walls were some of the features of the buildings that did not meet the design requirements.

During the months of April through June of 1988, approximately 6 months before the Spitak earthquake, 4 apartment buildings of series I-451 that had serious damage in load-bearing elements during construction were investigated by employees of Arm NIISa in Leninakan. Results of the investigations indicated that these buildings were constructed with substantial deviations from the actual design plans. Wall connections were not reinforced, seismic belts were missing, voids in brickwork were found, and junctions between panels were not filled. The low quality of construction, combined with the unauthorized construction of large openings in load-bearing walls by the inhabitants, caused serious damage to the load-bearing stone walls, exfoliation of brickwork, crushing and falling out of bricks, and cracks near apertures. In addition, the grout used in building construction contained many chemicals in the sand that caused an undesirable interaction with the cement resulting in very low strength. Damage was more severe within the first story of the buildings.

Study of the same buildings after the earthquake revealed 3rd and 4th degree of damage to the majority of buildings. In many buildings, severe damage occurred to the exterior walls and corners, especially in the upper stories.

There are a few examples of good construction. Buildings that were constructed without serious violations of the design plans and with consideration of seismic requirements performed relatively well during the earthquake. This confirms the idea that it is possible to construct stone buildings that are able to withstand strong earthquake shaking.

6. Large Panel 9-Story Buildings of Series A1-451 KP-16/1

Buildings of this type (see Figure 5) typically have large balconies on each side that protrude as far as 2.5 meters. The transverse walls of the balconies are a continuation of the main walls of the building and are included in the structural system for resisting seismic loads. The outer walls are made of one layer of light concrete 300 mm thick. The inner walls have round cavities of 140 mm in diameter that are placed 200 mm from each other. The partitions are made of pre-stressed concrete plates with many cavities. The buildings have strip foundations of heavy monolithic concrete. Despite some deficiencies in the design of certain elements, series A1-451 KP 16/1 buildings are generally designed in accordance with the standards for seismic resistant construction.

A study of all 16 large-panel 9-story series A1-451 KP 16/1 buildings in Leninakan showed that they performed satisfactorily in the Spitak earthquake. The average degree of damage to the buildings was 0 to 1. Small thin oblique cracks of less than 0.4 mm in width were observed in partition walls and in horizontal and vertical wall joints. Theoretical analysis of the earthquake performance of the 9-story large-panel buildings showed their high resistance to seismic forces. In contrast, all 19 of the 9- and 12-story series 111 frame-panel buildings located in the same area either collapsed or were so heavily damaged that they were later demolished.

7. 10- and 16-Story Residential Buildings Constructed by Lift Slab Method

These 10- and 16-story residential buildings were constructed for a maximum seismic intensity of VII. The 10-story building was designed in 1967 and had the shape of two round towers. The towers are the center of rigidity, have a diameter of 7 meters, have 18 cm thick flooring plates without girders, and are made of monolithic ferrous concrete surrounded by several columns. The flooring plates are welded to the columns, which are long enough to reach two floors. The design of the 10-story residential building was

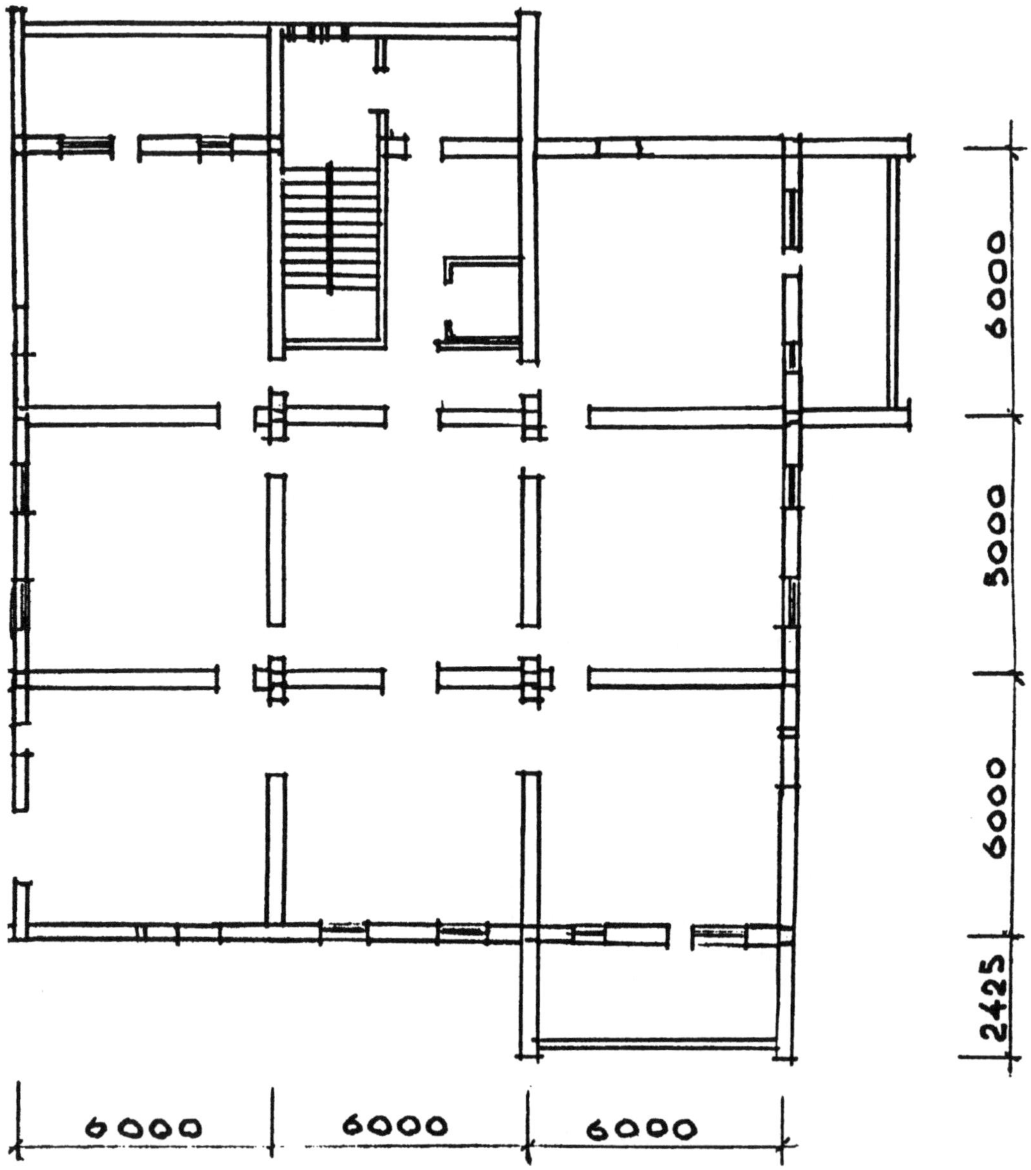

Figure 5. Large-panel 9-story residential building of series A1-451 KP 16/1

developed on the basis of scientific studies that later became the groundwork for "Temporary Regulations for Designing Civil Buildings Constructed by Method of Floorings and Stories Lifted" (SN 451-72). The 16-story building was designed in 1973 and has the shape of one rigid tower that looks like a polygon on the outside and a circle on the inside.

In both designs, some considerations that affect the seismic performance of these buildings were omitted. For example, adjacent elements of inter-story floors were not supported by the centers of rigidity resulting in lower earthquake resistance of the centers of rigidity. The construction system of the lift-slab method was never dynamically tested before the Spitak earthquake. There are no experimental results describing the behavior of this system. The only structural element of the building that is designed for seismic loads is the center of rigidity, which is a statically determinate system with no redundancy in the case of failure of one or more components.

During the Spitak earthquake, one 10-story lift-slab building was completely destroyed and one 16-story lift-slab building sustained heavy torsional deformations in the lower stories and had to be demolished. Patterns of damage in the 10-story building indicated that it also suffered from torsional deformations before collapse. Investigations carried out by the state commission experts in relation to SNiP P-7-81 showed that in the 10- and 16-story lift-slab buildings, torsional moments were several times larger than those specified in the designs. Another mistake was that the design seismic intensity was reduced from VIII to VII due to the relatively good soil in the area of the building foundations. It was also discovered that there were serious violations of the design requirements during the construction of these two buildings. For example, the concrete strength in the centers of rigidity was 1.3 to 1.7 times less strong than specified in the design documents.

The slightly better performance of the 16-story lift-slab building in comparison to the 10-story lift-slab building can be explained by the spectral content of the Spitak strong ground motion. The predominant period of motion appeared to be closer to the predominant period of the 10-story building (about 0.9 seconds) than that of the 16-story building (about 1.5 seconds).

8. Summary of Damage in Leninakan During the 1988 Spitak Earthquake

Out of a total number of 583 4- to 5-story and 9- to 12-story typical residential buildings in Leninakan, the Spitak earthquake caused partial or full destruction to 195. Two hundred thirty-one of the buildings sustained such damage that they had to be demolished, and 126 of the buildings were fully restored. Thirty-one of the buildings performed satisfactorily during the earthquake, including 15 4- to 5-story buildings and all 16 large-panel 9-story buildings.

In general, the combination of the following negative factors led to the catastrophic earthquake damage:

- The strength of the seismic impact was more than specified in the design of the buildings
- The characteristics of the earthquake were detrimental, including the strong aftershock, the coincidence of the spectral content of the motions with the dynamic properties of the buildings, and the strong vertical component of shaking
- The violation of design regulations and the lack of observance of the main principles of seismic resistant construction
- The low quality of construction

The low quality of construction in combination with the violations in design regulations played a major role in the grave consequences of the earthquake, especially in Leninakan where the seismic intensity level exceeded the design value by at least I.

After the earthquake, all residential, social, and industrial buildings and other structures in the most devastated areas were inspected. A detailed description of the performance of residential buildings was carried out and helped to develop projects for the strengthening of various types of buildings. Technical solutions for strengthening of residential buildings have changed since 1969. They have been improved to account for the better understanding of the earthquake performance of various types of building construction. In addition to traditional methods for strengthening buildings, several new experimental methods have been used on buildings that were not heavily damaged in the earthquake. These include base isolation and the addition of a separated or flexible top story.

Projects have been organized in Armenia to improve the technical basis for seismic resistant design and construction standards and regulations considering the experience of the 1988 Spitak earthquake. The first national standard for seismic resistant construction, SNRA II-2.02-94 "Seismic Construction - Design Norms," was developed and introduced into practice. Starting in October of 1996, new standards for design, construction, and inspection of stone and reinforced stone structures, SNRA W13.01-96 "Stone and Reinforced Stone Structures," were implemented. A new system of control of construction quality and an appropriate basis for design and construction standards and regulations has been created in Armenia. It is hoped that these new developments will also be applied to the construction of new individual houses, which is occurring rapidly.

Figure 6. Damage to side section of residential house of series 1A-450.

Figure 7. Damage to side section of residential house of series 1A-450.

Figure 8. Damage to side wall of residential house of series I-451.

Figure 9. Damage to parapet and side wall of residential house of series I-451.

THE DECEMBER 7, 1988 SPITAK, ARMENIA EARTHQUAKE: RESULTS OF ANALYSIS OF STRUCTURAL BEHAVIOR

V. RZHEVSKY
Multidisciplinary Center for Earthquake Engineering Research (MCEER), USA

1. Introduction

On December 7, 1988 at 11:41 AM local time, a strong earthquake with severe consequences occurred in the Republic of Armenia. The epicenter of the earthquake was in the Lesser Caucasus Highlands, about 80 km south of the ridge of the main range of the Caucasus Mountains. Seismographs all over the world recorded the Armenia earthquake. The event parameters reported by the Institute of Physics of the Earth of the USSR were a magnitude of 7.0 and a depth of 10 km.

The earthquake named "Spitaksky" occurred over a vast territory. It affected a population of about 700,000 people. The towns of Leninakan (population about 250,000) and Spitak (about 25,000) were completely destroyed. Heavy damage and failures of structures were observed in Kirovakan town (population about 170,000). Villages located in the epicentral zone of the earthquake were also completely destroyed (about 60 were leveled and 100 were damaged). The earthquake resulted in a loss of life estimated at about 25,000, and about 12,000 people were hospitalized. The economy of the entire region was affected as a result of this earthquake. Economic loss totaled more than sixteen billion dollars.

2. Background on Armenia

Armenia is situated in the southern region of Europe in part of the former USSR known as the Caucas area. Most of the Armenian territory is in the Caucas Highlands. The population at the time of the earthquake was about 3.6 million people. The main cities of Armenia included Yerevan (population about 1.0 million), Leninakan (population about 250,000), and Kirovakan (population about 170,000). Spitak was a small town with a population of about 25,000 people.

This entire region has a high degree of seismic activity. The long-term seismicity of Armenia indicates that earthquakes have damaged Yerevan and Leninakan in the past, especially in 1926, when many buildings in Leninakan were destroyed. Close to this region a very strong earthquake with magnitude of 8.0 occurred in 1939 in Arzrum,

S.A. King et al. (eds.),
Seismic Hazard and Building Vulnerability in Post-Soviet Central Asian Republics, 197–227.

Turkey. In 1983, just before the Spitak earthquake, a strong earthquake of magnitude 7.2 occurred at a distance of only 150 km from Leninakan. Both of these earthquakes occurred in the territory of other countries and were not considered in the assessment of seismic hazard in the Armenian region. In addition, in 1981 the seismicity of Armenia was reduced from intensity VIII to VII in the territory encompassing the towns of Kirovakan and Spitak. At the time of the earthquake, all of Armenia was divided into two seismic zones. A maximum seismic intensity of VIII was assigned to Yerevan and Leninakan, and intensity VII was assigned to Kirovakan and Spitak. According to the new map of seismic zoning, part of the territory of Leninakan was assigned a maximum seismic intensity of VII. The seismicity of Armenia has been described here in detail because it is important for the analysis of building behavior under seismic loads.

In Armenia only a few types of structural systems were used. They are 3- to 5-story stone-masonry buildings, pre-cast reinforced concrete frame-panel buildings, and pre-cast reinforced concrete large-panel 5- and 9-story buildings. Most of the industrial buildings were 1-story pre-cast reinforced concrete frames. All structures were designed to a seismic intensity level of VII to VIII. Many factors should be taken into account when evaluating the devastating effects of the Spitak earthquake. These include characteristics of ground motions, quality of design, quality of structures, type of building construction, soil characteristics, and position of buildings relative to the direction of seismic waves. It appears that the earthquake intensity, design quality, and quality of construction were the most important factors affecting earthquake damage. These are the factors that have been emphasized during analyses of the earthquake and its effects.

3. Observations of Building Performance: Assessment of Intensity

It was difficult to define the zone of the most intensive shaking, but it was observed that the most heavily damaged buildings were in Spitak, Leninakan, Kirovakan, and a number of settlements in the vicinity of these towns. The town of Spitak and the settlements of Nalband, Shenovan, and Sheracamut were in the epicentral zone of the earthquake. Soil rupture, rock falls, massive damage to roads and railways, and pipeline breaks were observed here. All these features correspond to a very high intensity of earthquake shaking.

Several examples of building performance in the epicentral zone of the earthquake are shown in Figures 1 through 5. Figure 1 presents a panoramic view of one of the districts of Spitak. It can be seen that all construction was completely destroyed. Almost all modern buildings were destroyed in Spitak. Figure 2 shows 5-story masonry residential buildings that collapsed in the earthquake. Nearly all of the residential houses in the settlements, located in the epicentral zone of the earthquake were completely destroyed. In most cases the quality of masonry was poor, and these houses were subjected to earthquake shaking of very high seismic intensity.

Figure 1. Panoramic view of Spitak, Armenia following the 1988 earthquake.

Figure 2. Collapsed 5-story residential masonry building in Spitak.

Figure 3. Damage to 5-story masonry buildings in Kirovokan.

Figure 4. Damage to residential masonry buildings in Leninakan.

Figure 5. Damage to Magnitoprovod plant industrial pre-cast frame building in Leninakan.

Stone masonry bearing-wall buildings were completely destroyed in Spitak, about 90% of existing buildings collapsed or had to be demolished. In Leninakan, about 50% of the buildings were heavily damaged. All buildings of this type were destroyed in one of the districts of Kirovakan. Examples of stone masonry bearing-wall building damage are shown in Figures 3 and 4. Complete destruction of industrial frame buildings, especially those with pre-cast reinforced concrete frames, was observed in Leninakan, Spitak, and Kirovakan. Although the seismic intensity of the earthquake was high, poor quality of construction contributed to these building failures.

In addition to the damaged buildings, there were structures that preformed rather well in the earthquake. There were two pre-cast large-panel buildings in Spitak that were not heavily damaged. These buildings withstood the earthquake shaking relatively well. Minor and moderate damage was observed in the panel connections, and there was a great deal of nonstructural damage. All 16 9-story pre-cast large-panel buildings in Leninakan also survived the earthquake without any significant damage. Only small cracks were observed between the wall panels.

On the basis of all building performance information and macroseismic analysis, it can be concluded that the intensity of seismic shaking in the epicentral zone was IX, corresponding to a peak horizontal acceleration of 0.4g. In Leninakan the seismic intensity was VIII to IX, corresponding to accelerations of 0.2g to 0.4g, and in Kirovakan the intensity was VII to VIII with accelerations of 0.1g to 0.2g. According to the report of the Soviet Post-Earthquake Investigation Team, "... the surface intensity of the earthquake in the area of the epicenter was not less than X according to the MSK-64 Intensity Scale."

4. Seismological Characteristics of the Earthquake

Instrumental recordings of strong ground motion during the Spitak, Armenia earthquake were obtained only at the Ghukasian seismic station, located at an epicentral distance of about 25 km. These records were of high quality and provide the opportunity to evaluate, although at only one location, the seismological characteristics of the earthquake. Visual inspection alone gives much information about the earthquake.

Two aspects of this earthquake are important. First, two, not one, strong earthquakes occurred in a four-minute interval. The analysis of building should take the occurrence of the two events into consideration. The buildings were subjected to a very strong earthquake and then were immediately subjected to a second high intensity earthquake. In some cases the buildings that survived the first earthquake collapsed in the second event. The second aspect is that the strong motion portion of the first earthquake had a relatively long duration of about 20 seconds. Earthquakes of this magnitude typically have a 10 to 12 second duration of strong shaking.

The characteristic parameters of the Ghukasian strong motion recordings were determined by statistical processing of peak acceleration values, and are shown in Table

1. Recorded peak ground acceleration values were 0.22g in the first earthquake and 0.15g in the second event. These acceleration values correspond to a seismic intensity of VII ½ for the first earthquake and VI ½ for the second event. Based on these records, the peak horizontal acceleration in Leninakan was estimated at 0.4g.

TABLE 1. 1988 Spitak Earthquake. Parameters of Recorded Accelerations at Ghukasian

Accelerogram	Maximum spectra of reactions, K_W ,sec	Duration, n , sec	Standard, σ_0 cm/s2	$\ddot{A}_{max}$, cm/s^2	Normalized coefficient T_{en}	
					A_a =0.2g	A_a =0.4g
Ghukasian 1, N - S	0.2 - 0.5	15.0	35.9	196.5	1.40	2.80
Ghukasian 1, E - W	0.15 - 0.6	14.6	34.0	221.0	1.45	2.90
Ghukasian 1, Z	0.1 - 0.9	11.6	23.6	144.3	1.70	3.40
Ghukasian 2, N - S	0.35	4.7	33.0	150.0	3.55	7.1
Ghukasian 2, N - S	0.2 - 0.8	7.2	20.8	104.4	3.65	7.3
Ghukasian 2, N - S	0.15 - 0.25	9.4	10.3	48.6	4.40	8.8

The frequency characteristics of the earthquake accelerations were determined by response spectra analysis and are shown in Table 1 and Figure 6. It can be seen that the predominant periods of horizontal shaking are in the range of 0.2 to 0.6 seconds. It should be noted that the shaking from the second earthquake in the east-west direction was characterized by a wider frequency range of 0.2 to 0.8 seconds. The range of predominant periods for the vertical component of shaking was unusually wide at 0.1 to 0.9 seconds. In general, vertical components of earthquake shaking typically have a narrow band in the high frequency range.

For the analysis of building behavior, it is important to know not only the maximum spectral values of seismic motion, but also the change in spectral characteristics during the duration of earthquake shaking. These characteristics are very important for the Spitak earthquake because most of the buildings were severely damaged and many were completely destroyed. The stiffness and other characteristics of buildings were changing as the earthquake shaking occurred. For this reason, it is necessary to estimate the 'frequency-time' properties of the earthquake. The results of spectral analysis of the Ghukasian accelerograms show that as the earthquake (the main shock) shaking progressed, the spectral characteristics (i.e., periods of vibration) of the motions increased. This had a negative effect on the behavior of buildings because as they became damaged, their natural periods also increased to the range of the seismic motions resulting in resonance problems.

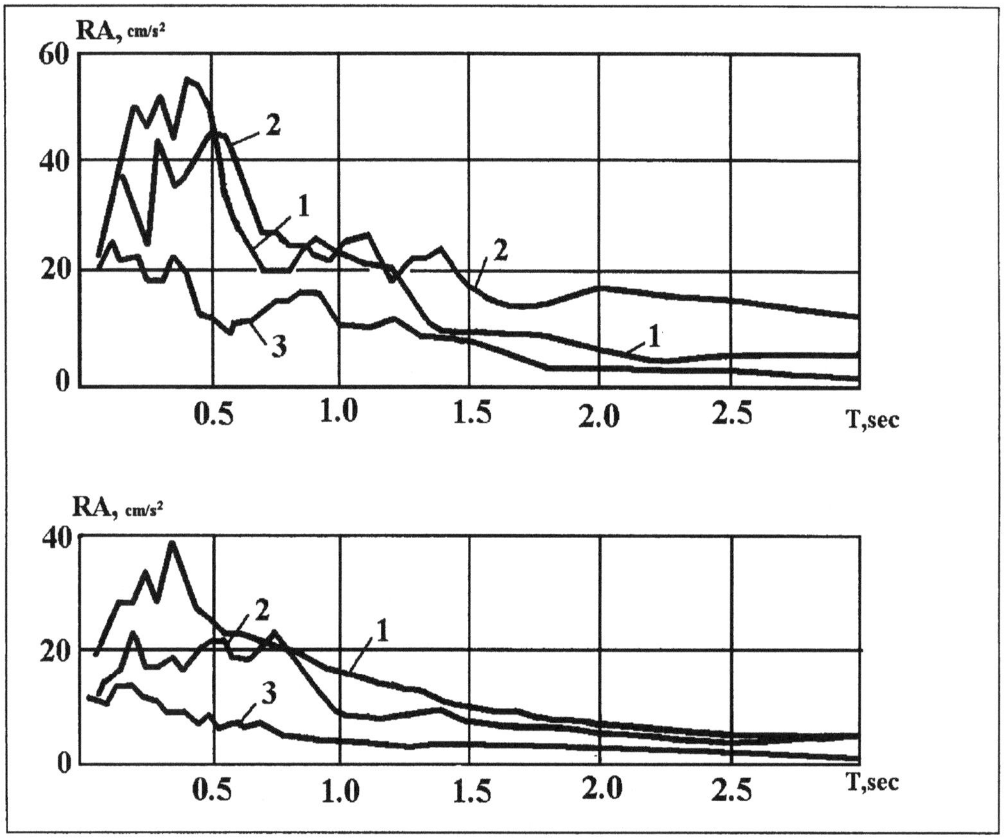

Figure 6. Response spectra of Ghukasian recordings of the 1988 Spitak earthquake.

5. Nine-story Precast Reinforced Concrete Frame-Panel Buildings of the 111 Series

The tragic consequences of the Spitak earthquake are primarily associated with the collapse of mass-produced modern buildings, particularly the 9-story buildings of the 111 series. These buildings were designed on the basis of the EES-04 series of frame-panel buildings. The frames are intended to be moment resisting frames constructed from pre-cast reinforced linear elements (i.e., beams and columns). The elements are assembled in place by welding in the zones of maximum forces, bending moments in this case. Such construction does not allow for the development of plastic deformations and results in drastically reducing the seismic resistance of the building.

There are additional problems with the EES-04 frames in the 111 series buildings.

In the transverse direction the buildings have rigid diaphragms, but in the longitudinal direction the structural system functions as a frame. This design leads to a reduction in the overall space rigidity of the building, and as a result to severe torsional vibration problems. The results of analyses have shown that the design flaws had a negative effect on these buildings during the earthquake. In general, the behavior of these buildings was very bad, especially in Leninakan where 95% of the pre-cast reinforced concrete frame buildings collapsed or required demolition. All 9-story buildings in Leninakan were torn down.

Because the devastation was so great, detailed analysis of possible reasons for massive collapse of these buildings has been done. For such analysis it is necessary to consider not only elastic, but also inelastic structural behavior. Analyses were conducted using nonlinear non-stationary models with the basement level subjected to actual earthquake loading. The Ghukasian records of acceleration were used as the external dynamic input.

Several variations in design were analyzed. The primary analysis was done for a 9-story building, designed for a maximum intensity of VIII, the design intensity of the damaged buildings. Load-bearing capacity, in terms of shear forces at the story levels, was determined by actual structural element and joint strength. The analysis was carried out separately for the transverse and longitudinal directions. Dynamic characteristics were determined according to the actual rigidity of the structural elements. Story masses were determined according to the standard design of the building. The fundamental periods were 0.87 seconds in the transverse direction and 1.34 seconds in the longitudinal direction.

The analysis was conducted as follows:

1. The building was subjected to the accelerogram of the first earthquake, and the damage of the system was determined
2. If the building did not collapse and it had reserve bearing capacity, the building was subjected to the accelerogram of the second earthquake; the initial data for analysis with the second earthquake were the strength and rigidity parameters of the building after the first earthquake
3. If the building collapsed, the time at which the collapse occurred was determined

The analyses were conducted with the computer program RUPS. The parameters that were computed at each time step, and for every floor, included reactions, accelerations, absolute and relative remaining deformation capacities, drifts, and energy of inelastic deformations. Changes in the building rigidity and in the fundamental period of vibration of the building were also determined. An example is given in Figure 7, which shows the characteristics of story deformation as a function of time. Figure 7 also shows the energy accumulation of the structure in the plastic range of deformation in both the transverse and longitudinal directions. The results of the analysis are given in Tables 2 and 3, where the maximum values of the computed parameters and the limit conditions are shown. Comparison of the computed parameters with the limiting values allow the seismic resistance of the buildings to be determined.

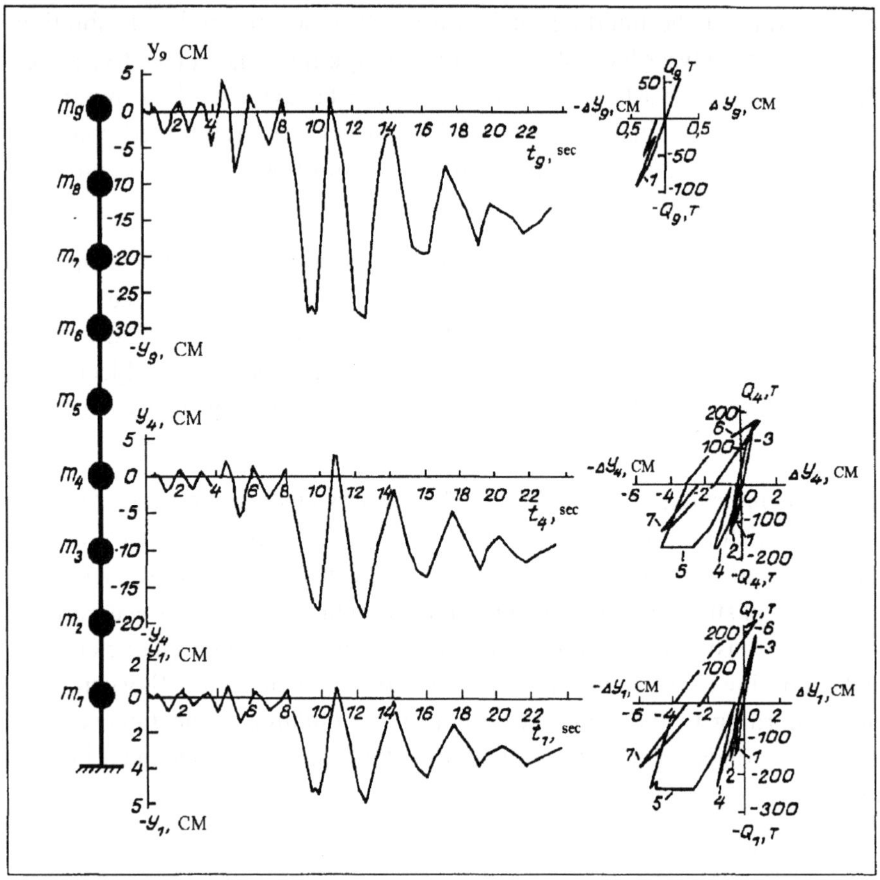

Figure 7. Analysis results of transverse direction of 9-story pre-cast reinforced concrete frame-panel building of 111 series subjected to accelerogram of Ghukasian (north-south; intensity 8, 0.2g).

The analysis results show that the 9-story frame buildings could not withstand the earthquake shaking in the areas where the intensity of seismic motion was IX. In these zones the buildings were destroyed in both the transverse and longitudinal directions. The buildings were destroyed in the first few plastic cycles, at 6 to 7 seconds into the earthquake shaking. All 9-story buildings in Spitak exhibited this behavior, and presumably most of the buildings in Leninakan did as well.

In the region where the earthquake intensity was VIII, very large deformations developed in both the transverse and longitudinal directions of the buildings. The analysis shows that in order to provide building safety, the displacements and story drifts had to exceed maximum permissible values by 30 to 60% at the first and second stories in the transverse direction and by 20 to 40% at the second, fourth, and eighth stories in the longitudinal direction. Such deformations will cause collapse of even well designed and well built reinforced concrete monolithic structures.

TABLE 2. Results of Analysis of 9-story Series 111 Buildings Subjected to Ghukasian 1 Accelerogram (transverse direction, T = 0.87 seconds)

Floor	Variant 1, N-S, A_a = 0.4 g					Variant 2, E-W, A_a = 0.4 g				
	K_W	$\bar{Y}_k$	$\Delta\bar{Y}_k$	n	T_{en}	K_W	$\bar{Y}_k$	$\Delta\bar{Y}_k$	n	T_{en}
1	39.7	12.7	12.7	6		219.0	62.0	62.0	10	
4	19.4	9.4	6.8	5	1.84	87.0	43.4	26.0	8	2.45
7	6.0	6.7	3.0	5		2.3	24.9	1.9	3	
9	0	4.7	0.6	0		9	17.0	0.4	0	
Floor	Variant 5, N-S, A_a = 0.2 g					Variant 6, E-W, A_a = 0.2 g				
1	14.2	6.3	6.3	5		44.6	16.7	16.7	4	
4	7.6	5.1	4.1	5	1.58	2.7	8.2	2.6	4	1.57
7	0.1	3.6	1.1	1		1	5.2	0.9	0	
9	0	2.9	0.5	0		1	4.1	0.3	0	
Floor	Variant 9, N-S, A_a = 0.15 g					Variant 10, E-W, A_a = 0.15 g				
1	8.1	3.8	3.8	5		14.8	6.7	6.7	3	
4	2.8	2.8	2.0	4	1.33	0.7	3.2	1.7	1	1.21
7	0	2.0	0.9	0		0	2.2	0.7	0	
9	0	1.6	0.4	0		0	1.8	0.3	0	

TABLE 3. Results of Analysis of 9-story Series 111 Buildings Subjected to Ghukasian 1 Accelerogram (longitudinal direction, T = 1.34 seconds)

Floor	Variant 3, N-S, A_a = 0.4 g					Variant 4, E-W, A_a = 0.4 g				
	K_W	$\bar{Y}_k$	$\Delta\bar{Y}_k$	n	T_{en}	K_W	$\bar{Y}_k$	$\Delta\bar{Y}_k$	n	T_{en}
1	17.8	7.2	7.2	5		83.1	22.1	22.1	7	
4	21.6	6.3	7.3	6	2.58	106.0	20.1	26.3	10	3.85
7	35.0	5.9	8.7	9		121.0	19.2	27.8	12	
9	3.5	5.7	2.5	2		12.2	18.8	5.0	3	
Floor	Variant 5, N-S, A_a = 0.2 g					Variant 8, E-W, A_a = 0.2 g				
1	4.4	2.9	2.9	2		31.2	11.0	11.0	7	
4	6.7	2.4	3.1	2	1.98	39.5	9.5	12.0	7	3.04
7	13.3	2.3	3.7	4		45.3	9.0	13.6	8	
9	0.8	2.2	1.4	2		1.2	8.2	2.0	1	
Floor	Variant 11, N-S, A_a = 0.15 g					Variant 12, E-W, A_a = 0.15 g				
1	2.8	1.6	1.6	2		19.0	5.2	5.2	3	
4	3.2	1.4	1.5	1	1.56	21.0	5.0	5.4	3	2.30
7	8.5	1.4	1.8	2		25.0	4.6	5.0	3	
9	0	1.5	0.3	0		0	3.2	0.6	0	

Due to the fact that the design and construction of frame-panel buildings of 111 series do not provide load-bearing capacity in the plastic range of deformation, it is concluded that seismic resistance of these buildings is not adequate for earthquakes of intensity VIII. Analysis results show that VII is the maximum seismic intensity that these buildings can withstand. All 9-story buildings in Kirovakan, where the intensity of earthquake shaking was about VII, performed satisfactorily.

6. Earthquake Resistance of 9-Story Buildings, Designed According to Building Code II-7-81

The performance of 9-story buildings in the Spitak earthquake was very poor. As analysis results show, it is primarily due to the inadequate quality of the design of these buildings. It is interesting to determine how these buildings, designed in accordance with the current building code, would behave under conditions similar to the Spitak earthquake. To study this problem, an analysis was done of a 9-story reinforced concrete frame-panel building with the same overall dimensions, mass, and rigidity as the buildings of the 111 series. The system was designed in accordance with the building code of the USSR for a design seismic intensity of VIII.

Results of this analysis are shown in Table 4. The analysis results show that earthquake resistance of this building is not provided for the Spitak earthquake level IX intensity. During the earthquake motion, inelastic deformations develop very quickly and the collapse occurs 12 seconds after the shaking has started. The maximum drift at the first floor level in the longitudinal direction is 10 cm. Energy reserves are depleted on all levels in the longitudinal direction and on the first through fourth floors in the transverse direction. It was also found that at level VIII intensity of the Spitak earthquake, plastic deformations develop in the system, but they do not exceed limiting values. For example in the transverse direction, maximum drift is 60 to 70% of the permissible limit on the first five to six stories. The upper stories behave elastically. Thus, seismic resistance of 9-story reinforced concrete buildings, designed with the building code, was adequate for the regions where the intensity of seismic motions did not exceed VIII (the design level). In this case seismic resistance of the building depends primarily on the structure's ability to sustain plastic deformations.

7. Evaluation of the Seismic Resistance of Reinforced Concrete Frame Industrial Buildings

At the time of the Spitak earthquake, most industrial buildings that were located in the epicentral zone were heavily damaged or collapsed. In general, these were one-story reinforced concrete frame buildings. A few actual buildings were analyzed after the earthquake.

TABLE 4a. Results of Analysis of 9-story Code Buildings Subjected to Ghukasian 1 Accelerogram (transverse direction, T = 0.87 seconds)

Floor	Variant 1, N-S, A_a = 0.4 g					Variant 2, E-W, A_a = 0.4 g				
	K_W	$\bar{Y}_k$	$\Delta\bar{Y}_k$	n	T_{en}	K_W	$\bar{Y}_k$	$\Delta\bar{Y}_k$	n	T_{en}
1	11.2	3.8	3.8	5		83.7	26.6	26.6	2	
4	35.7	4.7	7.5	6	1.75	67.9	20.3	18.5	9	2.19
7	4.5	4.5	2.4	5		1.2	14.5	1.5	3	
9	0	3.6	0.7	0		0	11.7	0.5	0	
Floor	Variant 5, N-S, A_a = 0.2 g					Variant 6, E-W, A_a = 0.2 g				
1	4.7	2.9	2.9	4		10.0	5.5	5.5	2	
4	10.0	3.1	4.3	6	1.42	9.2	4.4	5.1	4	1.30
7	0.5	2.8	1.2	2		0	3.3	1.0	1	
9	0	2.3	0.5	0		0	2.7	0.4	0	

TABLE 4b. Results of Analysis of 9-story Code Buildings Subjected to Ghukasian 1 Accelerogram (longitudinal direction, T = 1.34 seconds)

Floor	Variant 3, N-S, A_a = 0.4 g					Variant 4, E-W, A_a = 0.4 g				
	K_W	$\bar{Y}_k$	$\Delta\bar{Y}_k$	n	T_{en}	K_W	$\bar{Y}_k$	$\Delta\bar{Y}_k$	n	T_{en}
1	14.0	6.1	6.1	4		63.8	17.6	17.6	7	
4	16.1	5.9	6.2	4	2.56	81.7	18.5	21.4	9	3.81
7	26.0	6.0	7.2	8		93.8	19.0	22.9	11	
9	2.7	5.6	2.3	2		9.7	18.1	4.3	3	
Floor	Variant 7, N-S, A_a = 0.2 g					Variant 8, E-W, A_a = 0.2 g				
1	3.0	2.5	2.5	2		22.5	8.2	8.2	6	
4	3.9	2.3	2.4	2	1.96	27.9	8.3	9.3	6	2.98
7	10.1	2.2	3.1	4		32.7	8.5	10.5	6	
9	0.5	2.2	1.2	2		1.0	7.7	1.8	1	

7.1 LABORATORY BUILDING OF THE IGIS INSTITUTE IN LENINAKAN

The building is a one-story one-span (18 meters) reinforced concrete moment resisting frame with steel diagonal braces in the longitudinal direction. The building is equipped with a travelling crane. Because of the crane, the building design was based on a two-mass cantilever discrete model with masses concentrated at the level of the beams supporting the crane and the roof. The height of the building is 10.8 meters.

The building was heavily damaged but did not collapse. The analysis was conducted in the same manner as the analysis for the 9-story buildings. It was found that the fundamental period of vibration of the building was 1.06 seconds in the transverse direction and 0.88 seconds in the longitudinal direction. The results of the analysis are shown in Figure 8 and Table 5. The results show that the building, under earthquake shaking of intensity IX, should collapse in the longitudinal direction. In reality, this building survived the Spitak earthquake in the zone with intensity IX seismic motions.

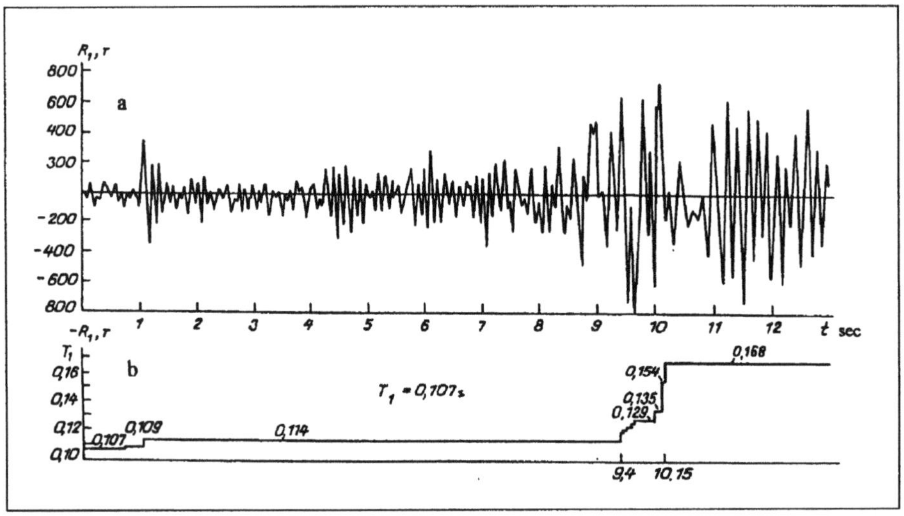

Figure 8. Seismic vibration of transverse direction of 5-story masonry building subjected to Ghukasian record (north-south, intensity 8, 0.2g). Predominant period of vibration is 0.11 to 0.17 seconds.

The analysis shows that under certain conditions the building might be heavily damaged but not collapse. It appears that the majority of the seismic forces acted on the building in the transverse or east-west direction, while in the longitudinal direction the diagonal braces collapsed due to the earthquake forces. In the analysis the collapse occurred early in the shaking motion, at about 3 to 4 seconds. These conditions would lead to a drastic change in the rigidity of the building. In this case the fundamental period increases from 0.88 to 2.12 seconds, which is no longer in the range of the predominant periods of the seismic motions. To verify this behavior, the building was analyzed in the longitudinal direction for the case when the frames are functioning without diagonal braces. In this case the inelastic deformations developed, but they did not exceed the limiting values.

TABLE 5a. Results of Analysis of IGIS Industrial Frame Building (transverse direction, T = 1.06 seconds)

Accelerogram	Number of story	Variant	K_{W_k}	$\overline{Y}_k$	$\overline{\Delta Y}_k$	n	T_{end} sec
Ghukasian, E-W, int. 9	1	1	6.1	2.5	2.5	4	2.13
	2		59.0	3.8	10.2	11	
Ghukasian, N-S, int. 9	1	3	30.0	9.6	9.6	6	2.71
	2		123.9	12.1	23.6	10	
Ghukasian, E-W, int. 8	1	5	2.4	2.2	2.2	4	1.80
	2		28.8	3.7	10.1	5	
Ghukasian, N-S, int. 8	1	7	7.6	4.0	4.0	3	2.0
	2		40.9	5.4	12.4	8	
Ghukasian, E-W, int. 7.5	1	9	1.2	1.4	1.4	3	1.63
	2		19.7	2.4	7.1	6	
Ghukasian, N-S, int. 7.5	1	11	1.8	2.2	2.2	1	1.65
	2		19.1	2.9	7.6	7	

TABLE 5b. Results of Analysis of IGIS Industrial Frame Building (longitudinal direction, T = 0.88 seconds)

Accelerogram	Number of story	Variant	K_{W_k}	$\bar{Y}_k$	$\Delta\bar{Y}_k$	n	T_{end} sec
Ghukasian,	1	2	13.5	4.5	4.5	7	2.14
E-W, int. 9	2		44.4	6.3	8.6	12	
Ghukasian,	1	4	45.0	13.7	13.7	6	2.59
N-S, int. 9	2		106.9	20.5	23.8	8	
Ghukasian,	1	6	4.1	2.1	2.1	4	1.82
E-W, int. 8	2		22.7	5.5	7.0	6	
Ghukasian,	1	8	10.3	5.6	5.6	3	1.93
N-S, int. 8	2		31.6	8.5	10.4	9	
Ghukasian,	1	10	2.1	1.7	1.7	4	1.66
E-W, int. 7.5	2		13.9	4.1	5.2	6	
Ghukasian,	1	12	6.4	3.7	3.7	1	1.65
N-S, int. 7.5	2		16.6	6.2	7.5	5	

The relatively good performance of the laboratory building of IGIS in Leninakan can probably be explained by the following two factors:

- The favorable orientation of the structural system
- The actual intensity of the seismic motion may have been less than IX

It is clear that the distribution of the diagonal braces played a positive role in preventing the collapse of the building in the longitudinal direction. After the braces failed in the early part of the shaking, the building was still able to adequately resist the earthquake forces.

7.2 BUILDING OF THE MAGNITOPROVOD PLANT

The structural system of this building was a pre-cast reinforced concrete one-story multi-span frame. The length of the spans was 18 meters and columns were spaced at 6 meters. Columns were 6 meters tall and 40 cm by 40 cm in cross-section. The building completely collapsed in the Spitak earthquake. Buildings of this type were widely used throughout Armenia and were heavily damaged or collapsed in various areas subjected to different levels of earthquake forces. For this reason, analysis of the behavior of this system was carried out for various intensity levels of earthquake shaking.

The method of analysis was almost the same as was used for the IGIS frame building described in the previous section. However in this case, it was assumed that the rigidity characteristics of the building were equal in both the longitudinal and transverse directions, so it was not necessary to analyze both directions. The fundamental period of these buildings is taken to be 1.62 seconds for buildings with height equal to 6 meters, and 1.15 seconds for buildings with height equal to 4.8 meters.

The results of the analysis are given in Table 6. The results show that the structural system of this building would never be able to withstand earthquake shaking of intensity IX. The actual deformations of the building were much more than the permissible limits for reinforced concrete structures. In regions with intensity of VIII, the buildings were also destroyed. Only in regions where the earthquake intensity did not exceed VII ½ did these types of buildings withstand the Spitak earthquake, but even in this case, the plastic deformations were very close to the limit values.

It can be concluded that the primary reason for heavy damage to one-story frame buildings was the high intensity of seismic shaking. The secondary reason has to do with the relatively low seismic resistance of one-story frame buildings. A reduction of 0.8 in the factor of safety for a frame building was required by the seismic design code of the former USSR. The analysis shows that without such decreases in the factor of safety, the one-story pre-cast reinforced concrete frames could have withstood the Spitak earthquake.

TABLE 6. Results of Analysis of 1-Story Magnitoprovod Industrial Frame Building in Leninakan

	T_b = 1.62 sec						T_b = 1.15 sec					
Accelerogram	Variant	K_W	$\bar{Y}$	$\sum \alpha$	n	T_{end}	Variant	K_W	$\bar{Y}$	$\sum \alpha$	n	T_{end}
Ghukasian, W-E, int. 9	1	39.7	28.7	35.2	6	4.70	7	40.5	10.7	16.7	7	2.79
Ghukasian, N-S, int. 9	2	128	26.2	47.9	10	5.08	8	157	32.7	58.9	10	3.82
Ghukasian, W-E, int. 8	3	9.6	5.5	3.7	1	2.69	9	15.9	4.9	6.8	7	2.23
Ghukasian, N-S, int. 8	4	47.4	13.3	19.0	6	4.03	10	63.3	17.7	26.7	8	3.13
Ghukasian, W-E,int. 7.5	5	5.4	3.4	2.4	1	2.42	11	9.3	3.5	4.5	4	2.01
Ghukasian, N-S, int. 7.5	6	25.4	8.1	10.8	5	3.50	12	33.6	9.7	14.8	6	2.70

8. Five-Story Composite Frame-Stone Masonry Buildings

One of the most widely used types of buildings in Armenia was the 5-story residential building of the 1-450 series. In the transverse direction these buildings have load-bearing walls spaced at 6.1 and 3.1 meters. In the longitudinal direction seismic loads are carried by two external walls and one internal wall. The external walls have many openings, and the internal walls are partly replaced by reinforced concrete frames. In general, the structural system of these buildings can be classified as a composite stone masonry load-bearing wall system with reinforced concrete elements. The disadvantages of this system were identified after the 1976 Gazli earthquake. This system was not recommended for use in seismic regions according to the design code of the former USSR.

The performance of these buildings in the Spitak earthquake was very poor. Many of them were severely damaged and collapsed. Ninety percent of these buildings collapsed or were later demolished in Spitak, and 60% were destroyed in Leninakan. These buildings were heavily damaged even in regions where the intensity of seismic shaking was VII or less.

An analysis was carried out to study the earthquake performance of the 5-story buildings. The analysis was done with a specialized computer program based on the CSIB (Controlling System with Inverse Braces) design model. This model allows for the representation of the properties of the earthquake accelerogram and the actual properties of the structural system when subjected to the seismic loading.

Stability and rigidity characteristics of the entire building are determined on the basis of the stability and rigidity parameters of individual structural elements (e.g., walls, columns, beams, and joints). At every time step of the earthquake motion, the demand on each element is determined depending on the current rigidity of the element. The demands are compared with the load-carrying capacity of the elements. If at any time step an element is damaged resulting in a change in the rigidity and stability of the element, the properties of the entire system are changed and the dynamic parameters of the building are recomputed. The analysis continues in this manner of updating system parameters according to damaged elements until the end of the strong ground motion record. The structural system characteristics are changed as each element is damaged during the earthquake. At each stage, or occurrence of damage, new strength and dynamic properties are calculated for the building. The model is a non-stationary system that changes its parameters according to the damage of the structural elements.

The 5-story buildings were analyzed for the entire range of seismic intensities of the earthquake shaking. In each case, damaged elements and the type of damage (e.g., shear force, torsion, tension, compression, and bending moment) were determined. The stiffness degradation of the structural elements and the entire building and the fundamental periods of vibration were also determined. Results of the analysis are presented in Figure 8, which shows the change in the fundamental period of the building as a function of the acceleration response time history.

According to this analysis, at a seismic intensity of IX all columns and beams in the east-west direction were damaged causing the building to collapse. It was also determined that the period of the first mode of vibration of the building increased by a factor of two. Nearly the same results were observed in the north-south direction, as 80% of the beams and 90% of the columns were destroyed. The results did not show much better building performance in the region with level VIII seismic intensity. In this case all 16 piers of the first story sustained severe damage causing building collapse. The analysis showed that the 5-story masonry buildings provided life safety only in the regions where seismic intensity was not more than VII.

9. Lift-Slab Buildings

Lift-slab building construction is widespread in Armenia and is now beginning in other seismic regions of the former USSR. These buildings have one or more cast-in-place reinforced concrete cores and pre-cast reinforced concrete columns for supporting the slabs. Slabs are cast on the ground level and lifted into their positions at the appropriate floor levels. These buildings are designed on the assumption that all lateral forces are resisted by the core. The columns function only to support vertical loads.

At the time of the Spitak earthquake, there were only two lift-slab buildings located in Leninakan. About 100 lift-slab buildings were located in Yerevan, which is fortunately located about 100 km from the epicenter of the earthquake. Both lift-slab buildings in Leninakan were completely destroyed. The 10-story building collapsed, and the 16-story building was so damaged that it had to be demolished.

The two destroyed buildings were analyzed to assess their seismic resistance and determine the reasons for failure. The analysis was conducted based on non-stationary inelastic models subjected to earthquake motions of various levels of seismic intensity. The results of the analysis, in terms of deformation and strength parameters, are shown in Tables 7 through 10.

The 10-story building collapsed at intensity IX shaking because maximum deformations of the central core were more than 10 times greater than the elastic deformations. The amplitudes of displacement were about 30 cm at the top of the building and 7 to 8 cm at the first floor level. In the case of level VIII intensity of shaking, analysis results show the story drifts to be 3 to 5 times greater than the values for elastic deformation. The amplitudes of displacement were about 13 cm at the top of the building and 3 cm at the first floor level. It is assumed that these values are beyond the allowable limit of deflections for this type of building.

For the 16-story building, the analysis shows that seismic resistance is provided for intensity VIII shaking. The building behaves mostly in the elastic range. Plastic deformation appears only at the bottom of the building and is greater than the elastic deformation values by only 10 to 20%. Seismic resistance of the building is provided for intensity IX shaking in the east-west direction, but limit values for deformation are reached in the north-south direction. The story drift values at the first through eighth floor levels are 2 to 3 times the limiting values of deformation for a rigid core.

The greater seismic resistance of the 16-story building relative to the 10-story one can be explained by the dynamic characteristics of the Spitak earthquake motion. The predominant periods of shaking were closer to the predominant period of vibration of the 10-story building (estimated at 0.9 seconds) than the 16-story building (estimated at 1.5 seconds). The 10-story building was in resonance with the earthquake shaking for a longer period of time. The damage to the core of the building began to develop in the early phase of the earthquake shaking, in the first few seconds, and the building collapse occurred about 10 to 12 seconds after the shaking had started.

10. Primary Reasons for the Severe Damage in the Spitak Earthquake

The following are the primary reasons for the severe damage that occurred as a result of the Spitak earthquake:

- Analysis of seismologic information has shown underestimation of seismic hazard for the territory of Armenia, as well as the whole Transcaucasian territory. The maximum possible intensity for this territory, according to the maps of seismic zoning, should not be more than VIII (acceleration equal to 0.2 g) in Leninakan and VII (acceleration equal to 0.1g) in the epicentral zone, including the town of Spitak. It should realized that an error in two levels of seismic intensity causes a decrease in the level of design seismic load by a factor of 4.
- Analysis of the seismic resistance of 9-story frame-panel buildings showed that the main reasons for their heavy damage and collapse are:
    - Seismological peculiarities of the earthquake (e.g., high intensity, frequency composition, long duration, and two strong earthquakes)
    - Low quality of design, resulting in the lack of reserves of load-bearing capacity in structural elements – In accordance with building code requirements, the structural connections should not be located in the zones of maximum stress, and they should allow for the development of plastic deformations; in pre-cast frames, consisting of single members, this requirement can not be satisfied because all beam-columns joints are in the locations of maximum bending moments (the welded joints do not guarantee the safety of the building)
    - Variations in the dynamic parameters of buildings in the transverse and longitudinal directions could lead to problems with torsional vibrations
- The main reason for heavy damage to one-story reinforced concrete industrial buildings was the high intensity of earthquake shaking. The secondary reason has to do with the lower seismic safety of these building relative to other structural systems.

TABLE 7. Results of Analysis of 10-Story Lift-Slab Building with Actual Strength (Ghukasian 1 Record; T = 0.93 seconds)

Floor	Variant 1, N-S, $A_a = 0.4$ g					Variant 2, E-W, $A_a = 0.4$ g				
	K_W	$\bar{Y}_k$	$\Delta\bar{Y}_k$	n	T_{en}	K_W	$\bar{Y}_k$	$\Delta\bar{Y}_k$	n	T_{en}
1	37.1	10.2	10.2	10		20.1	5.7	5.7	5	
4	18.3	8.6	7.3	4	1.93	14.2	5.5	4.9	5	1.76
7	5.1	6.7	3.4	3		4.9	4.3	2.2	4	
10	0	4.8	0.6	0		0	3.0	0.5	0	
Floor	Variant 5, N-S, $A_a = 0.2$ g					Variant 6, E-W, $A_a = 0.2$ g				
1	8.5	5.2	5.2	3		9.5	3.2	3.2	5	
4	3.8	4.0	3.2	2	1.35	3.6	2.9	2.3	4	1.39
7	0.1	2.9	1.1	1		0.1	2.1	1.1	1	
10	0	2.1	0.3	0		0	1.6	0.3	0	
Floor	Variant 9, N-S, $A_a = 0.15$ g					Variant 10, E-W, $A_a = 0.15$ g				
1	3.5	3.1	3.1	1		5.1	2.2	2.2	5	
4	0.6	2.3	1.5	1	1.12	1.5	2.0	1.4	3	1.21
7	0	1.6	0.7	0		0	1.4	0.9	0	
10	0	1.2	0.2	0		0	1.1	0.3	0	

TABLE 8. Results of Analysis of 10-Story Lift-Slab Building with Code Strength (Ghukasian 1 Record)

Floor	Variant 3, N-S, A_a = 0.4 g					Variant 4, E-W, A_a = 0.4 g				
	K_W	$\bar{Y}_k$	$\Delta\bar{Y}_k$	n	T_{en}	K_W	$\bar{Y}_k$	$\Delta\bar{Y}_k$	n	T_{en}
1	171.5	35.8	35.8	10		54.4	9.6	9.6	12	
4	111.6	31.0	25.6	9	2.91	40.6	9.2	8.1	11	2.28
7	80.7	26.8	21.0	7		31.5	8.2	7.0	8	
10	2.5	21.5	2.4	1		0.9	6.3	1.6	2	
Floor	Variant 5, N-S, A_a = 0.2 g					Variant 8, E-W, A_a = 0.2 g				
1	61.4	15.6	15.6	10		23.6	5.6	5.6	7	
4	36.8	12.8	9.9	9	2.23	18.9	5.2	5.1	6	1.91
7	22.8	10.7	8.0	5		19.3	4.9	5.3	6	
10	0.2	8.4	1.2	1		0	4.0	1.0	0	
Floor	Variant 11, N-S, A_a = 0.15 g					Variant 12, E-W, A_a = 0.15 g				
1	30.3	9.8	9.8	8		15.8	4.4	4.4	6	
4	16.5	8.3	7.0	4	1.87	13.1	5.0	5.4	5	1.77
7	11.0	6.9	5.3	4		12.4	4.7	4.8	6	
10	0	5.5	0.7	0		0	3.7	0.7	0	

TABLE 9. Results of Analysis of 16-Story Lift-Slab Building with Actual Strength (Ghukasian 1 Record; T = 1.50 seconds)

Floor	Variant 1, N-S, A_a = 0.4 g					Variant 2, E-W, A_a = 0.4 g				
	K_W	$\bar{Y}_k$	$\Delta\bar{Y}_k$	n	T_{en}	K_W	$\bar{Y}_k$	$\Delta\bar{Y}_k$	n	T_{en}
1	8.3	3.4	3.4	3		1.0	1.5	1.5	2	
4	4.7	2.9	2.5	2		0.4	1.3	1.2	2	
8	1.7	2.3	1.8	2	2.14	0	0.9	1.0	1	1.58
12	0	1.9	1.0	1		0	0.9	0.9	0	
16	0	1.6	0.2	0		0	0.8	0.2	0	
Floor	Variant 5, N-S, A_a = 0.2 g					Variant 6, E-W, A_a = 0.2 g				
1	0.6	1.2	1.2	2		0	0.7	0.7	0	
4	0.3	1.1	1.1	2		0	0.6	0.6	0	
8	0	1.0	0.9	0	1.55	0	0.5	0.5	0	1.50
12	0	0.9	0.6	0		0	0.5	0.5	0	
16	0	0.8	0.1	0		0	0.4	0.1	0	

TABLE 10. Results of Analysis of 16-Story Lift-Slab Building with Code Strength (Ghukasian 1 Record)

Floor	Variant 3, N-S, A_a = 0.4 g					Variant 4, E-W, A_a = 0.4 g				
	K_W	$\bar{Y}_k$	$\Delta\bar{Y}_k$	n	T_{en}	K_W	$\bar{Y}_k$	$\Delta\bar{Y}_k$	n	T_{en}
1	46.5	14.3	14.3	6		7.6	4.5	4.5	2	
4	34.4	13.1	11.5	6		6.2	4.0	3.6	1	
8	22.9	11.9	8.2	5	3.61	13.3	3.6	3.0	6	2.38
12	27.0	11.6	9.2	4		4.8	3.6	2.6	3	
16	0	9.8	0.9	0		0	3.0	0.6	0	
Floor	Variant 7, N-S, A_a = 0.2 g					Variant 8, E-W, A_a = 0.2 g				
1	11.3	4.9	4.9	4		1.6	2.1	2.1	2	
4	7.1	4.1	3.5	4		1.0	1.3	1.7	2	
8	3.7	3.7	2.6	3	2.54	0.7	1.5	1.3	2	1.85
12	2.6	3.5	2.2	3		1.5	1.4	1.7	2	
16	0	3.0	0.4	0		0	1.3	0.4	0	

- The main reasons for heavy damage to stone masonry buildings are the high intensity of seismic motions, the violations of building regulations for seismic areas, and the low quality of building materials.
- The earthquake illustrated the high seismic resistance of pre-cast concrete large-panel buildings due to the peculiarities of this structural system. These buildings only experienced slight damage in the earthquake.

11. Conclusions

In evaluating the consequences of the Spitak earthquake as a whole, it appears that the main reasons for the poor behavior of structures include:

- Incorrect evaluation of seismic hazard resulting in the buildings being designed for too low levels of seismic forces – Even well designed and well built structures typically can not withstand seismic shaking of intensity two to four times greater than the design intensity
- Very poor quality of construction work
- The use of structural systems that did not provide reserves of load-bearing capacity in the inelastic range of deformation – Without these reserves all structural systems (reinforced concrete frame-panel buildings, lift-slab buildings, and stone masonry buildings) could not provide life safety when subjected to seismic activity

The Spitak earthquake has shown that it is necessary to use physical methods for calculation, design, and analysis of seismic reliability and safety of existing and new buildings. In addition, the earthquake has shown, as has been demonstrated by several other earthquakes, that the behavior of structures is often influenced more by the frequency characteristics of the seismic motion than by the intensity of shaking. For this reason, frequency parameters of probable earthquakes should be considered in the seismic zonation process.

12. References

1. Abdurashidov, K.S., Vedernikov, A.A., Echmetov, L.X., Babich, L.I., and Abdukarimov, Z. (1989) Engineering Analysis of the Spitak Earthquake Consequences, *Building and Architecture of Uzbekistan,* 12-14.
2. Rashidov, T.R., Xozhmetov, G.X., Ishanxodzhaev, A.A., Kasimov, X.K., Usupov, A., Fasakhov, V.G., Kornienko, V.P., and Raxmanov, U.E. (1989) Spitak Earthquake of 1988, *Building and Architecture of Uzbekistan,* 4-7.
3. Rzhevsky, V.A. and Avanesov, G.A. (1979) Limit State of Reinforced Concrete Elements and Frames, *Concrete and Reinforced Concrete,* 17-28.

4. Rzhevsky, V.A. (1981) Elastic-Plastic Properties of Reinforced Concrete Frame Systems, *Building and Architecture of Uzbekistan,* 6-10.

5. Rzhevsky, V.A. and Ibragimov, R.S. (1985) Building Analysis Using Earthquake Accelerograms with Allowance for the Damage of Load-Bearing Members, *Structural Mechanics and Structural Design,* 46-50.

6. Rzhevsky, V.A., Tsipenuik, I.F., Asamov, X.A., Gamburg, U.A., Uzlov, S.T., Khakimov, S.A., Muchamedchin, A.A., and Shirin, V.V. (1989) Engineering Analysis of Spitak Earthquake Consequences, *Building and Architecture of Uzbekistan,* 7-12.

7. Rzhevsky, V.A., Ibragimov, R.S., and Filiavich, V.N. (1989) Analysis of 5-Story Masonry Buildings in the Conditions of Spitak Earthquake, *Building and Architecture of Uzbekistan,* 15-16.

8. Rzhevsky, V.A., Ibragimov, R.S., Novoselia, A.V., Peschansky, L.V., and Rendel, P.A. (1989) Analysis of 9-Story Reinforced Concrete Frame-Panel Buildings in the Conditions of Spitak Earthquake, *Building and Architecture of Uzbekistan,* 27-30.

9. Rzhevsky, V.A., Ibragimov, R.S., Novoselia, A.V., Peschansky, L.V., and Suchkova, M.U. (1989) Estimation of the Seismic Resistance of Commercial Buildings in the Conditions of Spitak Earthquake, *Building and Architecture of Uzbekistan,* 30-33.

10. Rzhevsky, V.A. and Khakimov, S.A. (1990) Analysis of Lift-Slab Building Behavior in the Conditions of Spitak Earthquake, *Building and Architecture of Uzbekistan,* 11-13.

11. Rzhevsky, V.A. (1990) The Main Reasons of Heavy Consequences of Spitak Earthquake of 7/12/1988, *Building and Architecture of Uzbekistan,* 13-15.

12. Rzhevsky, V.A. (1990) *Seismic Resistance of Buildings Subjected to Strong Earthquake Motions*, Science, Tashkent.

13. Shirin, V.V. (1989) Damage of Masonry Buildings under Action of Spitak Earthquake of December 7, 1988, *Building and Architecture of Uzbekistan,* 33-37.

14. Tsipenuik, I.F., Gamburg, U.A., Gorbenko, V.M., and Musheev, M.D. (1989) Precast Large-Panel Buildings Behavior in the Conditions of Spitak Earthquake, *Building and Architecture of Uzbekistan,* 25-27.

15. Ulomov, V.I. (1989) Earthquake in Armenia: Element and Responsibility, *Building and Architecture of Uzbekistan,* 1-4.

LESSONS OF THE 1995 SAKHALIN AND 1994 KURIL ISLANDS EARTHQUAKES

A. MELENTYEV
Central Research Institute of Building Structures, Russia

1. Background on the 1995 Sakhalin Earthquake

On May 28, 1995 at 1:04 AM local time, a strong earthquake occurred to the north of Sakhalin Island. The MSK intensity was IX and higher in the vicinity of the town of Neftegorsk. A significant number of buildings were destroyed in Neftegorsk and 2000 inhabitants lost their lives. Adjacent settlements and the town of Okha, 40 miles to the north, were also damaged, although to a considerably lesser degree.

The settlement of Neftegorsk is located at a distance of 90 km to the south of the town of Okha. This relatively young settlement has been built since the 1960s in conjunction with the beginning of oil exploration. The oil and gas company "Vostokneftegaz," a member of the larger corporation "Sakhailinneftefaz," was located here. In the past few years, this company has decreased its capacity of oil output, and the settlement has become less populated.

The main construction of Neftegorsk was started in 1964. The buildings were constructed to last 30 years and include the following:

- 17 five-story large-block 80-flat residential buildings constructed from 1967 to 1971
- 2 two-story brick houses and 2 large-block houses constructed in the 1980s
- One-story three-family cottages and 4 two-story kindergartens constructed in the 1960s and 1970s
- Two-story buildings of the "Vostokneftegaz" administration, working shop, and the Palace of Culture for 360 persons
- Three-story 964-pupil brick school with a gymnasium constructed in 1969 and 1970
- Two-story buildings of the personal service shop and a dining house for 100 persons
- The polyclinic and 50-bed hospital with a dispensary

S.A. King et al. (eds.),
Seismic Hazard and Building Vulnerability in Post-Soviet Central Asian Republics, 229–235.

- Two-story buildings of a boiler-house, bakery, post-office, and working shops
- A number of wooden cottages

2. Building Characteristics and Damage in the 1995 Sakhalin Earthquake

Several circumstances had an effect on the type of damage to buildings and structures in Neftegorsk, including the proximity of the earthquake's epicenter to the town, the lack of seismic resistance considerations when buildings were constructed, depletion of oil deposits because of aging materials (claydite concrete blocks), and low quality of construction.

Five story large-block residential buildings were the most heavily damaged, as all 17 buildings totally collapsed. These buildings were not designed for construction in seismic, permafrost, and mining working conditions. The construction of these residences was carried out from 1964 to 1971 when this region was assumed to have a maximum seismic intensity of VI. Seismic resistance measures were not considered in the design. Dimensions of these buildings (series 1-447C-5/60) were 67.2 meters by 12 meters in plan with a height of 18.6 meters. Structurally, the buildings consisted of sections of large claydite concrete M75 blocks 400 mm thick with reinforced concrete round hollow cover plates. The internal transverse load-bearing walls were of brick 380 mm thick (of M75 bricks on M25 mortar). Foundations were of the strip type with sectional reinforced blocks 400 mm thick and a depth of roughly 4.5 meters. According to the data of laboratory tests, the real strength of the claydite concrete blocks was 26 kg/cm^2 instead of the design specification of 75 kg/cm^2.

Two-story large-block residential buildings of the 114-52-166C/1 and 113-123 series (constructed from 1986 to 1987) were located close to the destroyed five story buildings described in the previous paragraph. These two-story residences were constructed when Neftegorsk was located in a seismic zone of intensity VI according to the map of general seismic zoning in the Building Code II-7-81. The buildings were designed and built with some earthquake resistance measures, and they had longitudinal load-bearing walls spaced at 6.3 meters. These buildings are relatively rigid structures and sustained earthquake damage of the 1st and 2nd degrees. The specific damage included vertical and horizontal cracks between blocks and straight arches, diagonal cracks in partitions, damage to furnaces, vertical cracks in wall connections, partial damage to chimneys, and displacement of entrance canopies (one of them collapsed).

About 80 wooden one- and two-story cottages sustained no damage.

The high school building suffered heavy damage. It was comprised of two parallel sections connected by a passage. One section contained classrooms located in a three-story brick building. The walls of the building were of red bricks faced by silica bricks, and the spans were sectional reinforced concrete plates. The other section was a gymnasium, which was a large-span structure. In this building, reinforced concrete cover plates were supported on brick walls by reinforced concrete blocks. The distance between the axes of bearing walls was 12 meters. The two parallel sections were connected by a two-story passage with walls built from bricks. The roof was constructed of reinforced concrete plates and was supported on brick columns by reinforced concrete blocks. The school building was built without earthquake resistance measures and experienced damage to the 4th and 5th degree. The specific damage includes partial collapse of load-bearing walls, total collapse of a number of non-load-bearing walls, substantial damage to partitions (in some cases collapse), cracks between span plates, and deep diagonal cracks up to 20 mm thick in brick walls. The roof collapsed in the section where the auditorium was located. Total collapse (5th degree damage) occurred in the gymnasium building.

In the 1960s and 1970s, 4 two-story kindergartens were built in Neftegorsk (for 140 children). Walls of one building were of large claydite concrete blocks with reinforced concrete spans. The rest of the kindergarten buildings were frame structures with hinged panels. Building panels were 12 meters by 51 meters in plan with a height of 6.85 meters. The columns were 300 x 300 mm sections of reinforced concrete. The beams were reinforced concrete T-sections. The hinged panels were single layers of claydite concrete 320 mm thick. The spans were of round hollow reinforced concrete plates. The kindergarten buildings performed satisfactorily in the earthquake despite the lack of earthquake considerations in the design. Moderate damage was observed in the frame kindergarten buildings. Specific damage to these buildings included interlock cracks up to 4 mm wide in external load-bearing walls, vertical cracks in connections between transverse and longitudinal walls, exfoliation of plaster, and diagonal and x-shaped cracks in transverse walls and partitions.

The two-story polyclinic building with one-story add-on dispensary had load-bearing walls of large claydite concrete blocks. These buildings were constructed in the late 1960s and early 1970s without any design considerations for earthquake loading. The polyclinic building had damage to the 3rd degree, which included vertical and horizontal cracks up to 8 mm wide between blocks and straight arches in load-bearing and non-load-bearing walls, diagonal and x-shaped cracks in transverse walls and partitions, and separation of end transverse walls from longitudinal walls. The separation of longitudinal and transverse walls from each other was observed in the one-story dispensary addition.

The two-story frame building of the working shop collapsed during the earthquake (5th degree of damage). The shop was a reinforced concrete frame with hinged panels. The interior non-load-bearing walls and partitions were constructed in small cinder blocks of size 40x20x20 cm and M75 bricks. The building dimensions were 30.4 meters by 12 meters in plan.

Significant damage was observed in the frame building of the Palace of Culture, which consisted of two perpendicular sections. Total collapse occurred in the large-span section where the concert hall was located. The distance between columns was 12 meters and exterior walls were constructed of red brick. A discotheque was taking place in this hall during the earthquake resulting in many casualties. A frame building, which contained rooms of artistic amateur groups and the administration of the House of Culture was located adjacent to the concert hall building. Non-load-bearing walls and partitions were built of red bricks as in the concert hall. Damage to this building was of the 4th degree, and included deep diagonal cracks up to 3 cm wide in non-load-bearing walls, collapse of partitions, partial collapse of span plates at the ground floor level, and collapse of landings and staircases. Both of these sections were built without earthquake resistance design considerations.

3. Summary of Observations of 1995 Sakhalin Earthquake

The field investigations and post-earthquake studies of buildings and structures damaged in Neftegorsk by the Sakhalin earthquake of May 28, 1995, as well as the analysis of available data, have provided a large volume of information on how to improve the seismic design codes as well as the seismic safety of existing buildings and structures.

It was concluded that the main factors contributing to collapse of structures and human casualties in Neftegorsk are the following:

- The high degree to which the actual seismic loads on buildings exceeded the design values
- The high vertical component of earthquake acceleration combined with the low strength of walls
- The lack of seismic microzoning and the lack of pre-construction engineering and geological site investigations

Poor seismic behavior was demonstrated by concrete block wall buildings with low concrete strength and reinforced concrete frame buildings without diaphragms. Relatively good seismic behavior was observed in prefabricated large panel buildings and one-story wooden buildings.

4. Background on the 1994 Kuril Islands Earthquake

On October 4, 1994 at 4:22 PM Moscow time, a catastrophic earthquake occurred to the south of the Kuril Islands and 70 km east of Shikotan Island. The magnitude was 8.0 and the seismic intensity was IX to X. The main shock was accompanied by a tsunami and several aftershocks. During the period of October 4 through 19, 35 aftershocks occurred, including two with magnitudes greater than 7.0 and 9 with magnitude greater than 6.0.

5. Characteristics of buildings and earthquake damage on Shikotan Island

The type of building damage was influenced by weakening of the structures due to material aging (especially in wooden structures) and low construction quality.

Wooden panel-board houses, heated by claydite, were built in the early 1960s as temporary residences for workers of the fishing industry. No seismic design considerations were included. The foundations were concrete strip or concrete-wood footings. These buildings were damaged to the 3rd and 4th degree, and included the following types of damage:

- Deep vertical and slanted cracks n the foundations up to 1 cm wide
- Separation of the longitudinal walls from the transverse walls
- Collapse of the floor on the ground level because of the foundation displacement
- Damage to furnaces
- Collapse of chimneys
- Heavy damage to plaster of walls and ceilings

Wooden beam one- and two-story houses, built in the 1960s and 1970s, had damage to the 2nd and 3rd degree. The damage to the houses was mainly in the foundations and included deep cracks in the strip foundations and considerable displacements in wooden foundations up to 7 cm. In the building structures there was collapse of plaster from the walls and ceilings, sagging of the floors, cracks between longitudinal and transverse walls, damage to furnaces, and collapse of chimneys. It should be noted that the main cause of damage in these structures was the loss of bearing strength due to material aging. Rotted materials in the lower walls and foundation elements were observed in many houses.

In the stone buildings of Shikotan Island, 400x200x200 mm cinder blocks were used in the walls enclosing the structures. Several of the two-story houses with cinder block walls of the new construction in the 1980s have monolithic strip foundations, prefabricated overlaps, and seismic belts. The walls were constructed of plaster over lath. The degree of damage to these houses was 2 to 3. In general, the foundations were damaged by vertical and diagonal cracks, and the joints between transverse and longitudinal walls separated. Deep diagonal cracks were observed in transverse load-bearing walls, and partitions were severely damaged. The first stories were damaged much more than the ground floors.

Separation of the horizontal and vertical connections of panels, as well as collapse of entrance canopies in several houses, were the most typical types of damage to large-panel houses. The degree of damage to these buildings was 1 to 2. The type of damage observed in wooden buildings, with beam walls used for non-residential purposes, was the same as in wooden dwellings of the same type.

Severe damage was observed in the multi-story buildings with cinder block or brick walls and incomplete reinforced concrete framework. Uses for these buildings include school, restaurant, bakery, fire outpost, bath and laundry center, and storehouses for the fish-processing factory. The degree of damage to these buildings was 3 to 4. Typical damage included considerable damage to the foundations, diagonal cracks in the columns of longitudinal load-bearing walls, diagonal cracks in transverse non-load-bearing walls and partitions, and partial collapse of partitions, floor slabs, and end walls.

Frame buildings with brick or cinder block infill walls sustained the greatest damage. Almost all of them had damage to the 4th and 5th degree. The two-story school in the town of Malokurilsk was built of a monolithic reinforced concrete frame with brick infill walls. The school has two perpendicular sections, consisting of classrooms and a gymnasium. Damage to the brickwork of the infill walls was caused by deep diagonal cracks. Partial collapse of partitions was observed in the educational section. The framework of the ground floor had slanted cracks up to .5 mm near the beam-column connections. The gymnasium's non-load-bearing end walls were heavily damaged and partially collapsed.

Damage to the one-story frame buildings used as storage and containment facilities was similar to the school building damage. The club house in Malokurilsk was under construction at the time of the earthquake. It had an intricate configuration in plan and consisted of several separate sections. The first phase of this building, a gymnasium, was put into operation in 1993. The foundation was strip monolithic concrete, the load-bearing walls were cinder block, and the roof was constructed of reinforced concrete slabs on roof trusses. The roof totally collapsed during the earthquake causing damage to the 5th degree. The remainder of the club house had partial overlaps from pre-cast slabs that were lost during the earthquake. The degree of damage to this section of the building was 4. Separation of the hinged panels from the framework and the collapse of some slabs at overlaps were observed.

6. Summary of Observations of 1994 Kuril Islands Earthquake

The surveys of buildings damaged by the earthquake, as well as the engineering examination of available data, have given a considerable amount of information on necessary improvements to the seismic buildings code. Results of data investigations have confirmed the concepts of the current Russian seismic building codes, and also exposed a number of new facts which include the following:

- The high seismic resistance of large-panel buildings
- The relatively high seismic resistance of wooden buildings that are well designed and constructed
- The low seismic resistance of combined structural systems with a compound load-bearing structure, e.g., reinforced concrete frame with brick or stone infill walls –

Damage to these buildings was increased due to the fact that the buildings were designed without following the Russian seismic building code

- The substantial effect of local ground conditions on the number of structural failures

Other relatively new facts revealed by the building inspections after the earthquake include the following:

- This was the first time a base-isolated building experienced a strong earthquake in Russia; the seismic behavior of the base-isolated building was much better compared with the nearby non-base-isolated buildings
- The heavy damage of the foundations observed in this earthquake has not been observed in other large earthquakes

ASSESSMENT OF SEISMIC RESISTANCE OF SOVIET MASS CONSTRUCTION BUILDINGS - ALMATY AS AN EXAMPLE

T. ZHUNUSOV
KazNIISSA, Kazakhstan
I. ITSKOV
KazNIISSA, Kazakhstan
A. TAUBAEV
KazNIISSA, Kazakhstan

1. Introduction

The requirements for the design and construction of buildings in seismic regions are described in standard documents that include regulated methods for defining seismic loads and providing the minimum measures to ensure seismic resistance. All of the standard documents form a system of interconnected design and construction regulations that were developed based on instrumental data on the quantitative parameters of strong ground motion, results of engineering analysis, and results of experimental and theoretical research.

Based on research results and actual data on the earthquake performance of buildings with various types of designs, the opinions of experts on the adequacy and efficiency of some design schemes have changed over time. For example, in the last 50 years, design standards for construction in seismic zones of the former USSR were changed seven times. Currently at the institute KazNIISSA, a new version of the design standard is being developed for use in Kazakhstan. An analysis of how the design standards have been modified over time shows that the requirements for seismic resistance of buildings are getting higher. This tendency is similar to what has occurred in other countries. The most radical changes to design standards occur after a devastating earthquake produces many collapsed modern buildings in a large city. Engineering analysis of building performance in the earthquake allows the objective estimation of the adequacy and efficiency of the main principles of the current design standard and points out possible deficiencies.

Due to the constant modification in design standards, researchers and engineers often encounter problems associated with estimating the seismic resistance and design seismic load of buildings that were designed and constructed according to outdated standards. These problems become very apparent when dealing with buildings of standardized mass construction. An imperfection in the standard building design may

S.A. King et al. (eds.).
Seismic Hazard and Building Vulnerability in Post-Soviet Central Asian Republics, 237–244.

become the primary reason for devastating social and economic consequences due to a large earthquake.

Approximately 80% of the multi-story residential construction of Almaty, Kazakhstan consists of brick, frame, and large-panel buildings constructed according to standardized designs. Beginning in 1933, construction in Almaty was based on the document "Temporary Technical Conditions of Designing and Construction of Civil Objects in Seismic Regions in Kazakhstan." The standards for seismic resistant design, developed by the order of Kazsovnarcom and published in 1930, were at the time the most widely used in the former USSR. These standards were based on the standards used in Italy with little modification.

By 1935, changes and additions were made to the temporary design provisions, mainly to allow for the use of load-bearing brick walls in frame buildings. The new regulations were relatively severe and limited the height of brick buildings to three floors. Exterior walls at the ground floor were at least 3 bricks thick, at the first floor at least 2.5 bricks thick, and at the second and third floor at least 2 bricks thick. Columns and foundations were constructed of monolithic reinforced concrete. Several buildings constructed according to these regulations are still in use in Almaty at the present time. In the early 1930s, design for earthquake forces was based on the static theory. Design forces corresponded to a seismic intensity of IX and an acceleration of 100 cm/sec^2.

In 1957 in the USSR, "Norms and Regulations of Construction in Seismic Regions, SN 8-57" was developed. Seismic resistant design was based on dynamic theory. This standard has been revised many times. Currently, the seismic resistant design and construction in Kazakhstan is regulated by SNiP II-7-81*, "Construction in Seismic Zones." For design and construction in the territory of Almaty, the regional standards, developed in 1970 and modified twice, are used. The latest version, "Construction of Almaty and Surrounding Territories Taking into Account Seismic Microzoning" SN RK B.2.2-7-95, was developed by the institute KazNISSA and placed into operation in 1995.

2. Buildings with Load-bearing Brick Walls

Residential buildings with load-bearing brick walls of two or more stories were constructed in Almaty for a long period of time, but the intensive construction took place from 1930 to 1960. The construction of residential buildings of two or more stories with standardized designs in Almaty has stopped at the current time.

Brick buildings were designed according the standards that were in place during the various periods of construction. Because of this, the buildings have different capacities for resisting seismic loads. In addition, older buildings have deteriorated over time resulting in loss of strength. Residential brick buildings in Almaty can be divided into the following three groups:

1. Buildings with load-bearing walls of 2 or more stories with wooden columns – Most of these buildings were designed according to the static theory for seismic resistance and were constructed from 1930 to 1957
2. Buildings of four floors with columns of reinforced concrete – These buildings were constructed until 1957 according to individual designs
3. Buildings of four floors with prefabricated reinforced concrete slabs – Many of these buildings are of the standard series and were designed according to the dynamic theory of seismic resistance

The above grouping of residential brick buildings is relative and not intended to be detailed. It was made by the institute KazNIISSA when developing the document "Methodical Recommendations on Passport Systems of Present Construction Buildings in Almaty and other Settlements Situated in Seismic Hazards Zones of Kazak SSR" that was issued by Gosstroi of Kazak SSR in Almaty in 1989.

Buildings of the first group typically have low seismic resistance. The design of the brick settings was to a much lower standard than that given in SNiP II-7-81. Deficiencies in this type of buildings include:

- Insufficient stiffness of columns and inadequate anchoring of brick walls to wooden columns
- Rotting of wood in columns
- Existing cracks in brick walls due to soil subsidence and minor earthquake shaking of intensity VI or less

Buildings of this type are scheduled to be demolished. It is possible to strengthen them, but it is not economically feasible.

Buildings of the second group are typically able to withstand shaking of a relatively high intensity, but it is necessary to conduct an investigation of their design and construction to verify that they conform to the applicable standards for seismic resistance. The seismic resistance of the brick settings in these buildings typically satisfy the lower limit of seismic resistance requirements in the SNiP II-7-81 standard. It should be noted that the number of these buildings that was constructed is not large, and they are all typically built on soils with good engineering geology properties. A few have been built in fault zones, even locations with evidence of surface faulting.

The third group includes buildings of the standard series 275 and 308. These buildings are located in the central and southern parts of the territory of Almaty and have the following typical properties:

- Considerable mass
- Lack of a basement in part of the building
- Low seismic resistance of brick settings

- A minimum of seismic strengthening elements, primarily only seismic ties in columns
- Existing cracks in brick settings and in slab-column connections

When developing seismic strengthening measures for these buildings, it is necessary to analyze the design and construction to assess the current state of the seismic resistance of the building. Seismic strengthening methods should consider the requirements of the standard under which the buildings were designed, as well as the technical capabilities of local construction organizations. Some developments in this area have been done by the institute KazNIISSA, but it would be useful to create a special program for strengthening these buildings, including sources of funding for this program.

3. Large-Panel Buildings

In order to satisfy the housing requirements of the population of Almaty, construction of large-panel residential buildings made of pre-fabricated structural elements was undertaken. The majority of these buildings were constructed during the period of 1985 to 1988, when the amount of construction of large-panel buildings totaled about 400,000 m^2, which was roughly 70% of the total amount of residential construction in Almaty. Currently, the total area of large-panel buildings in Almaty and the surrounding region is about 9,700,000 m^2, and these buildings are occupied by 700,000 to 800,000 inhabitants.

Large-panel buildings in Almaty are typically of series 1-464-AC, 1-KZ464-DC, 69, E-147, and 158. Buildings of the 464 series were intended to be 4 to 5 stories tall, series 69 were 5 stories, series E-147 were 8 stories, and series 158 were 9 stories. The design of all large-panel buildings types was based on a transverse wall structural scheme that included longitudinal and transverse load-bearing walls. The walls were connected by vertical joints, and between floors they were assembled from compact slabs the size of one room. Four- and 5-story buildings had one interior longitudinal wall, while 8- and 9-story buildings had two.

The first large-panel buildings constructed in Almaty were 4-story buildings of series 1-464-AC. The spans of transverse walls were typically 2.6 meters and 3.2 meters, and the walls were assembled from three layer panels the size of one room. The outer panels were 250 mm thick, and the interior load-bearing panels were 100 mm thick. The interior walls were assembled from panels for one or two rooms that were 120 mm thick. The panels were reinforced by single steel mesh elements placed in the center of the panel. Joints between panels were made by welding steel and then filling with monolithic concrete.

The standard design of series 1-464-AC buildings was developed by the institute TsNIIEPzhilisha in Moscow according to the requirements of SN-8-57 and later the requirements of SNiP II-A 12-62. For seismic resistance of these large-panel buildings, flat design schemes were used. The interior longitudinal and transverse walls of the building were considered to be vertical elements, rigidly fixed at the ground floor level. The sizes of the vertical elements in plan corresponded to the sizes of continuous wall sections between door openings. The design of panels for the exterior walls assumed they acted as 1-story 1-bay frames with pinned connections. The overlaps were designed as rigid unstrained disks that supplied compatibility for the vertical elements.

In 1967, construction of large-panel buildings of series 1-464-AC stopped and construction of series 1KZ-464-DC buildings began in Almaty. The most significant differences between these two types of buildings are in the exterior wall panels and in the amount of reinforcement in the interior walls. The exterior wall panels on the series 1KZ-464-DC buildings had cylindrical cut-outs and reinforcing pins. The cut-outs provide a means for placing vertical reinforcing joints between adjacent panels throughout the entire height of the building. The reinforcing pins were welded and then the cut-outs were filled with concrete. The interior wall on the series 1KZ-464-DC buildings were reinforced by two meshes instead of one as in the series 1-464-AC buildings. Construction of the series 1KZ-464-DC buildings continued until 1992.

Large-panel building of series 69 (developed by institute Kazgorstroiproyect in 1973), series E-147 (developed by TsNIIEPzhilisha in 1972), and series 158 (developed by institute Almatygiprogor and TsNIIEPzhilisha in 1980) are referred to as the next generation of large-panel buildings. The level of comfort provided in these buildings, in terms of insulation for sound and heat, exceeded that provided in series 464 buildings. The spans of the transverse walls were 3.6 meters. The interior wall and ceiling panels were 160 mm thick, and the exterior wall panels were 350 mm thick. The joints between panels were considerably different in these buildings than buildings of the 464 series. Vertical joints between panels were made of welded reinforcing pins with concrete filling. Similar joints were used for connecting adjacent ceiling panels.

Seismic resistant design of series 69, E-147, and 158 buildings was done using design models, taking into consideration the spatial make-up of the structures and the results of experimental research. The first serious experiment on the seismic resistance of large-panel buildings in Almaty was carried out in 1967 while a mud-flow prevention dam was under construction in Medeo Canyon. Ground explosions necessary for the dam construction caused shaking intensity of VII to VIII within a 2 km radius of the construction site. Six full-scale models of buildings of different construction types were placed at a distance of 800 meters from the location of the explosions. All building models were constructed according to the design standards that were in force at the time. A maximum seismic intensity of IX was assumed for the seismic resistant design requirements.

The explosions at the Medeo Canyon dam construction site caused peak horizontal accelerations of 0.4g to 0.6g at the foundation level of the model buildings. The model buildings constructed of brick experienced peak horizontal accelerations of 0.9g in the upper stories and were severely damaged. The large-panel model buildings experienced peak horizontal accelerations of 0.6g in the upper stories and sustained only slight damage. The results of this experiment demonstrated for the first time the relatively high seismic resistance of large-panel buildings. The results also provided a good comparison between the seismic performance characteristics of brick and large-panel buildings.

Large-panel buildings may currently be considered the most experimentally studied structural systems in Central Asia. For the past 30 years, the institutes TsNIIEPzhilisha and KazNIISSA have conducted dynamic tests on approximately 50 full-scale large-panel model buildings. In addition, about 20 full-scale model buildings were tested by specialists from the institutes TsNIISK Kucherenko, TbilZNIIEP, ArmNIISSA, TashZNIIE, TISS, and others. Many of the model buildings were subjected to dynamic loads 2 to 3 times larger than the design seismic loads, and a few were loaded to capacity.

The adequacy of the seismic resistant design measures in large-panel buildings has not been proven in actual strong earthquake shaking. The most conclusive evidence of the relatively good seismic resistance of large-panel buildings was their performance in the 1986 Karakum and 1988 Spitak earthquakes. While these earthquakes caused many buildings collapses, there were no fatalities or serious injuries among the inhabitants of large-panel buildings. In addition, the condition of large-panel buildings was such that they could be occupied after these earthquakes.

Research on the dynamic behavior of large-panel buildings has shown that these buildings have a considerable amount of reserve ductility and the capacity for developing plastic deformations. In addition, it has been shown that overall structural systems and the connections between panel elements play very important roles in ensuring the seismic resistance of these buildings.

The problems with connections between panels are most apparent for large-panel buildings of series 1-464-AC, which were constructed by welding more than 30 years ago. In 1978 and 1990, while studying the seismic resistance of existing buildings and the ability to strengthen them, specialists from KazNIISSA conducted a detailed examination of two large-panel buildings of series 1-464-AC. These two buildings were constructed in 1960. The results of the examination showed that more than 30% of the joints connecting interior wall panels around staircases and near lavatories and kitchens were subjected to corrosion. The amount of remaining welding steel was 60% of the original amount. The other wall panel connections in the buildings, located in areas not subjected to moisture, were found to be in satisfactory condition.

Corrosion of connections between interior wall panels presents a serious hazard in large-panel buildings of series 1-464-AC. In order to provide life safety to the occupants of these buildings, it is necessary to examine the connections of all buildings and develop methods to strengthen those that are found to be corroded. This work was undertaken in 1989, but had to be stopped due to a lack of funding.

4. Frame buildings

Most recently, residential buildings in Almaty were constructed of prefabricated reinforced concrete elements or a combination of prefabricated reinforced concrete elements and cast-in-place concrete. Similar to other building types, the frame buildings were constructed according to standardized series designs, and to a maximum seismic intensity of IX. Practically all frame buildings were made of overlapping prefabricated reinforced concrete slabs with cast-in-place reinforced concrete connections. Several types of materials were used as infill walls in the frame buildings. Prefabricated light concrete panels were primarily used as infill walls in non-residential buildings.

The height of residential frame buildings varies from 4 to 9 stories. The majority of these buildings were constructed during the last 30 years and are assumed to have adequate seismic resistance according to the classification developed by KazNIISSA. The classification includes the following seismic resistant qualities:

- Limited number of floors in frame buildings
- Increased cross-sectional size of columns in ground floor and first floor levels
- Use of more reliable connections for hung panels

Reinforced concrete space frames were used in frame buildings of series VP-1 and VT-20. The column elements in these space frames were made of cast-in-place reinforced concrete and the beams were made of prefabricated reinforced concrete. All of the joints were constructed of welded reinforcing pins covered by concrete. In reality, many of these joints did not conform to the design requirements in terms of construction quality and durability. Frame buildings of series SzhKU-9 (5 to 9 stories) are considered to be more reliable than the series VP-1 and series VT-20 buildings because of the quality of the joint construction. The series SzhKU-9 buildings include transverse half-frame elements that have fully welded steel plate connections.

Prefabricated frames of series IIS-04-US and series IIS-04-3 (5 to 9 stories) were primarily used in the construction of administrative buildings in Almaty. Long reinforced concrete columns were joined by welding. Similarly, transverse and longitudinal beam joints, as well as beam-column connections, were made of welded steel. Many administrative buildings in Almaty of 5 to 9 stories were also constructed of series 1.020.1-2s frames. These frames are similar to those used in the infamous series 111 residential buildings in Spitak, Armenia that fully collapsed during the 1988 earthquake. Series 1.020.1-2s frames typically have insufficient space rigidity and poor construction quality in the welded joint connections.

A considerable number of kindergartens and schools in Almaty are 2-story buildings with series 2KZ-200S frames. These frames were fully prefabricated and connected by reinforced concrete slabs resting on the tops of columns. The structural design of these frames is not sufficient to withstand earthquake shaking of intensity VIII or IX. Full-scale model testing of a series 2KZ-200S frame by KazNIISSA showed that this type of frame has no reserve strength, and the stability and durability of the connections are not adequate for the design level of seismic loading. Currently, construction of this type of building in Almaty is forbidden, but the problem of how to strengthen existing buildings still remains. Methods for strengthening buildings with series 2KZ-200S frames have been developed by KazNIISSA and implemented in projects dealing with reconstruction of schools and kindergartens. This program has run out of funding and has been stopped.

5. Summary

It has been recommended that all existing buildings in Almaty be examined and classified into one of the following three groups:

- Those that conform to the current design and construction standards or that have adequate seismic resistance based on the examination results
- Those that were designed and constructed to the current standards, but that are seismically hazardous due to their existing structural condition
- Those that were not designed and constructed to the current seismic resistance requirements

The above classification is rather broad and does not contain enough information to be useful. While developing predictive maps of expected earthquake damage to buildings, specialists from KazNIISSA developed a method for classifying all existing buildings in Almaty according to their seismic resistance. The method is quite different from the classification scheme described above and includes the following steps:

- Assess the building's seismic resistance with a preliminary examination
- Determine if a detailed examination of the building is needed
- Develop alternatives for dealing with the current seismic resistance of the building, such as demolition, strengthening, or changing the use

In 1989, the steps listed above were improved and became the basis for the document "Methodical Recommendations on Making Passport System of Almaty Buildings and Other Settlements Situated in Seismically Hazardous Regions of Kazak SSR." By following the steps outlined here, all of the existing buildings in Almaty were divided into 13 categories. Category 1 buildings (with the highest seismic resistance) included large-panel buildings with monolithic joints and steel frame, and prefabricated frame buildings with large elements. The buildings with the lowest seismic resistance included brick building that have not been strengthened with reinforced concrete elements (category 10), and brick buildings with wooden frame elements (category 12).

APPENDIX: FORMS AND QUESTIONS FOR WORKSHOP PARTICIPANTS

Prior to the meeting in Almaty, the workshop organizers identified experts in the fields of seismology and structural engineering in each of the five Central Asian republics. These experts were invited to participate in the workshop. In addition to their participation, they were asked to submit a report describing the seismic hazard and building vulnerability in their republic, particularly the capital city. These reports are included in this book. To help them prepare their report, the workshop organizers sent them a series of forms and questions pertaining to seismic risk. The forms and questions are described in this Appendix.

Form 1: Main Information about Seismic Hazard and Buildings Vulnerability

1. Capital city's name (current one and any previous ones), geographic coordinates, and the year it was founded
2. Brief geomorphologic description of the capital city region
3. Seismo-tectonic description of the capital city, including a map if available
4. Map of the general seismic activity in the region, including magnitudes of past earthquakes
5. Size in area and population of the capital city
6. Amount of multi-story residential construction and the number of people who live in such buildings
7. Soil conditions in the capital city region and a map of seismic microzonation if available

Form 2: Characteristics of the Seismic Hazard and Expected Earthquake Intensity

1. Map of general seismic hazard zoning for the capital city region, with a city plan in the background if possible
2. Map of detailed seismic hazard zoning for the capital city region
3. Map of the seismic zones that may cause earthquakes in the capital city region

S.A. King et al. (eds.),
Seismic Hazard and Building Vulnerability in Post-Soviet Central Asian Republics, 245–249.

4. List of parameters for the main seismic zones in the region, including maximum magnitude, return periods for earthquakes of various magnitudes, distance to the capital city, and faulting mechanism
5. Expected earthquake magnitude and its return period according to the map of general seismic zoning for the USSR
6. Plot of recurrence relationship for expected magnitude in the capital city region
7. Plot of seismic intensity attenuation relationship for the region
8. Relationship for peak horizontal acceleration attenuation in the region, if there is one
9. If known, the expected ground motion parameters in the region, including acceleration, velocity, displacement, duration, and spectral content

Form 3: Destructive Earthquakes of the Past

1. Number of years that earthquake information has been collected
2. Number of earthquakes with magnitude 7 or larger that have occurred in the capital city region
3. For each large earthquake in the capital city region:
   - Parameters including date and time, geographic coordinates of epicenter, magnitude, distance from epicenter to capital city, faulting mechanism, population in capital city at the time, and number of casualties
   - Isoseismal map if possible
   - Main types of buildings at the time in the capital city
   - Percentage of heavily damaged buildings of each type
   - Brief description of damage to each building type
   - Possible reasons for damage to each building type
   - Secondary effects, such as landslide, liquefaction, and surface rupture

Form 4: Statistics on Different Residential Building Types of Mass Construction

1. List of the main residential building types for the capital city, which should be something similar to the following:
   - Brick buildings with longitudinal load-bearing walls
   - Brick buildings with transverse load-bearing walls
   - Brick buildings of other types
   - Frame buildings with brick walls
   - Frame buildings with cast-in-place elements
   - Frame buildings with precast elements

- Frame buildings with stiffness core
- Frame-panel buildings
- Large-panel buildings without stiffness core
- Large-panel buildings with stiffness core

2. For each building type, the following information:

   - Time periods of construction, including design code that was in force at the time
   - Number of people who live in buildings of such type

3. The most vulnerable building types, including the design scheme, series number, year(s) of construction, and the number of people who live in such buildings

Form 5: Scientific Research Institutes and Organizations of the Republic

1. Name and address of organizations in the capital city that develop national design and construction standards and regional regulations, and that enforce the requirements

2. Name and address of the main engineering organizations that carry out mass building design and construction in the capital city

Form 6: Regulations for Seismic Resistant Design and Construction

1. Name and valid time period of design and construction regulations over the past 40 years (time periods should correspond to list of buildings statistics on Form 4)

2. Brief description of seismic design considerations in each regulation

3. Description of how current building regulations are controlled and enforced

Form 7: Seismic Strengthening of Existing Buildings

1. Description of methods that are currently used for seismic strengthening of existing buildings in the capital city

2. Name of organizations that work on reinforcing residential buildings, and name of the design code that they use

3. Main obstacles to reinforcing the buildings that do not have adequate seismic resistance

4. Seismic design level to which the existing building are strengthened

Form 8: New Methods for Seismic Resistant Design

1. Name of organizations that are working on developing new methods for seismic resistant design
2. New documents that are being used in the design and construction of new buildings
3. Seismic design level to which the new buildings are being constructed
4. Number of new buildings, if any, that are being designed and constructed according to out-of-date and inadequate building regulations
5. Necessary measures to increase the seismic resistance of new residential buildings
6. Main obstacles faced by engineering firms that develop residential building projects in seismically active regions

Form 9: Optimal Method for Defining Seismic Loads

1. Information that design engineers would like to have from the seismologists in order to accurately define the seismic load on a building
2. Parameters other than seismic intensity that are important when defining the seismic load on a building, such as the following:
    - Maximum acceleration
    - Maximum velocity
    - Maximum displacement
    - Response spectrum
    - Plot of fundamental period
    - Time history of acceleration
    - Duration of shaking
3. Return period of the seismic load for which the building should be designed
4. Description of service, if there is one, for recording earthquake ground motion in buildings and in the free field
5. If applicable, the maximum seismic intensity and acceleration that have been recorded by this service

Form 10: Scientific and Technical Cooperation

1. Possible forms of collaboration among the five republics of Central Asia
2. Description of a permanent central coordinating center for information and activities concerning seismology and earthquake engineering in Central Asia

3. Themes and topics for future workshops on earthquake risk in Central Asia

4. Possible forms of cooperation with other countries (e.g., Unites States), such as the following:

    - Research projects
    - Design code development
    - Databases
    - Literature
    - Training of specialists

SUBJECT INDEX

C

D

E

F

G

MIX
Papier aus verantwortungsvollen Quellen
Paper from responsible sources
FSC® C105338
FSC
www.fsc.org

If you have any concerns about our products,
you can contact us on
ProductSafety@springernature.com

In case Publisher is established outside the EU,
the EU authorized representative is:
Springer Nature Customer Service Center GmbH
Europaplatz 3, 69115 Heidelberg, Germany

Printed by Libri Plureos GmbH
in Hamburg, Germany